TRANSPUTER REFERENCE MANUAL

INMOS Limited

Prentice Hall

New York   London   Toronto   Sydney   Tokyo

First published 1988 by
Prentice Hall International (UK) Ltd,
66 Wood Lane End, Hemel Hempstead,
Hertfordshire, HP2 4RG
A division of
Simon & Schuster International Group

© 1988 INMOS Limited

INMOS reserves the right to make changes in
specifications at any time and without notice. The
information furnished by INMOS in this publication is
believed to be accurate, however no responsibility is
assumed for its use, nor for any infringement of patents
or other rights of third parties resulting from its use. No
licence is granted under any patents, trademarks or
other rights of INMOS.

●, inmos, IMS and occam are trademarks
of the INMOS Group of Companies.

INMOS document number: 72 TRN 006 04

All rights reserved. No part of this publication may be
reproduced, stored in a retrieval system, or transmitted,
in any form, or by any means, electronic, mechanical,
photocopying, recording or otherwise, without the
prior permission, in writing, from the publisher.
For permission within the United States of America
contact Prentice Hall Inc., Englewood Cliffs, NJ 07632.

Printed and bound in Great Britain
at the University Press, Cambridge

CIP data are available

2 3 4 5 92 91 90 89 88

ISBN 0-13-929001-X

TRANSPUTER
REFERENCE MANUAL

Other titles in this series

Transputer Reference Manual
Transputer Development System
occam Technical Notes
Transputer Technical Notes
Transputer Instruction Set: a compiler writer's guide
Digital Signal Processing

Contents

	Preface		
	Notation and nomenclature		
1	**Transputer Architecture**		**1**
	1	Introduction	2
		1.1 Overview	3
		Transputers and occam	3
		1.2 System design rationale	4
		1.2.1 Programming	4
		1.2.2 Hardware	5
		1.2.3 Programmable components	5
		1.3 Systems architecture rationale	5
		1.3.1 Point to point communication links	5
		1.3.2 Local memory	6
		1.4 Communication	6
	2	occam model	8
		2.1 Overview	8
		2.2 occam overview	9
		2.2.1 Processes	9
		Assignment	9
		Input	9
		Output	9
		2.2.2 Constructions	10
		Sequence	10
		Parallel	10
		Communication	11
		Conditional	11
		Alternation	11
		Loop	12
		Selection	12
		Replication	12
		2.2.3 Types	13
		2.2.4 Declarations, arrays and subscripts	13
		2.2.5 Procedures	14
		2.2.6 Functions	14
		2.2.7 Expressions	14
		2.2.8 Timer	15
		2.2.9 Peripheral access	15
		2.3 Configuration	16
		PLACED PAR	16
		PRI PAR	16
		2.3.1 INMOS standard links	16
	3	Error handling	17

	4	Program development	18
	4.1	Logical behaviour	18
	4.2	Performance measurement	18
	4.3	Separate compilation of occam and other languages	18
	4.4	Memory map and placement	19
	5	Physical architecture	20
	5.1	INMOS serial links	20
		5.1.1 Overview	20
		5.1.2 Link electrical specification	20
	5.2	System services	20
		5.2.1 Powering up and down, running and stopping	20
		5.2.2 Clock distribution	21
	5.3	Bootstrapping from ROM or from a link	21
	5.4	Peripheral interfacing	21

2 Transputer Overview   23

	1	Introduction	25
	2	The transputer: basic architecture and concepts	26
	2.1	A programmable device	26
	2.2	occam	26
	2.3	VLSI technology	26
	2.4	Simplified processor with micro-coded scheduler	27
	2.5	Transputer products	27
	3	Transputer internal architecture	28
	3.1	Sequential processing	29
	3.2	Instructions	29
		3.2.1 Direct functions	30
		3.2.2 Prefix functions	30
		3.2.3 Indirect functions	31
		3.2.4 Efficiency of encoding	31
	3.3	Support for concurrency	31
	3.4	Communications	33
		3.4.1 Internal channel communication	33
		3.4.2 External channel communication	35
		3.4.3 Communication links	36
	3.5	Timer	37
	3.6	Alternative	37
	3.7	Floating point instructions	37
		3.7.1 Optimising use of the stack	38
		3.7.2 Concurrent operation of FPU and CPU	38
	3.8	Floating point unit design	39
	3.9	Floating point performance	39
	3.10	Graphics capability	41
		3.10.1 Example - drawing coloured text	41
	4	Conclusion	43

3		IMS T800 Engineering Data	45

	1		Introduction		46
	2		Pin designations		48
	3		Processor		49
		3.1	Registers		49
		3.2	Instructions		50
			3.2.1	Direct functions	50
			3.2.2	Prefix functions	50
			3.2.3	Indirect functions	51
			3.2.4	Expression evaluation	51
			3.2.5	Efficiency of encoding	51
		3.3	Processes and concurrency		52
		3.4	Priority		53
		3.5	Communications		54
		3.6	Timers		54
	4		Instruction set summary		56
		4.1	Descheduling points		57
		4.2	Error instructions		58
		4.3	Floating point errors		58
	5		Floating point unit		65
	6		System services		67
		6.1	Power		67
		6.2	CapPlus, CapMinus		67
		6.3	ClockIn		67
		6.4	ProcSpeedSelect0-2		68
		6.5	Reset		69
		6.6	Bootstrap		69
		6.7	Peek and poke		71
		6.8	Analyse		71
		6.9	Error, ErrorIn		72
	7		Memory		73
	8		External memory interface		75
		8.1	ProcClockOut		75
		8.2	Tstates		75
		8.3	Internal access		76
		8.4	MemAD2-31		77
		8.5	MemnotWrD0		77
		8.6	MemnotRfD1		77
		8.7	notMemRd		77
		8.8	notMemS0-4		77
		8.9	notMemWrB0-3		81
		8.10	MemConfig		84
			8.10.1	Internal configuration	84
			8.10.2	External configuration	85

		8.11	notMemRf	90
		8.12	MemWait	91
		8.13	MemReq, MemGranted	93
	9	Events		95
	10	Links		96
	11	Electrical specifications		99
		11.1	DC electrical characteristics	99
		11.2	Equivalent circuits	100
		11.3	AC timing characteristics	101
		11.4	Power rating	102
	12	Package specifications		104
		12.1	84 pin grid array package	104

4   IMS T414 Engineering Data   107

	1	Introduction		108
	2	Pin designations		110
	3	Processor		111
		3.1	Registers	111
		3.2	Instructions	112
			3.2.1 Direct functions	112
			3.2.2 Prefix functions	112
			3.2.3 Indirect functions	113
			3.2.4 Expression evaluation	113
			3.2.5 Efficiency of encoding	113
		3.3	Processes and concurrency	114
		3.4	Priority	115
		3.5	Communications	116
		3.6	Timers	116
	4	Instruction set summary		118
		4.1	Descheduling points	119
		4.2	Error instructions	119
	5	System services		124
		5.1	Power	124
		5.2	CapPlus, CapMinus	124
		5.3	ClockIn	124
		5.4	Reset	126
		5.5	Bootstrap	126
		5.6	Peek and poke	128
		5.7	Analyse	128
		5.8	Error	129
	6	Memory		130

	7	External memory interface		132
		7.1	ProcClockOut	132
		7.2	Tstates	132
		7.3	Internal access	133
		7.4	MemAD2-31	134
		7.5	MemnotWrD0	134
		7.6	MemnotRfD1	134
		7.7	notMemRd	134
		7.8	notMemS0-4	134
		7.9	notMemWrB0-3	138
		7.10	MemConfig	141
			7.10.1 Internal configuration	141
			7.10.2 External configuration	142
		7.11	notMemRf	147
		7.12	MemWait	148
		7.13	MemReq, MemGranted	150
	8	Events		152
	9	Links		153
	10	Electrical specifications		156
		10.1	DC electrical characteristics	156
		10.2	Equivalent circuits	157
		10.3	AC timing characteristics	158
		10.4	Power rating	159
	11	Package specifications		160
		11.1	84 pin grid array package	160
			11.1.1 84 pin PLCC J-bend package	162
5	IMS T212 Engineering Data			165
	1	Introduction		166
	2	Pin designations		168
	3	Processor		169
		3.1	Registers	169
		3.2	Instructions	170
			3.2.1 Direct functions	170
			3.2.2 Prefix functions	170
			3.2.3 Indirect functions	171
			3.2.4 Expression evaluation	171
			3.2.5 Efficiency of encoding	171
		3.3	Processes and concurrency	172
		3.4	Priority	173
		3.5	Communications	174
		3.6	Timers	174

4		Instruction set summary	176
	4.1	Descheduling points	177
	4.2	Error instructions	177
5		System services	182
	5.1	Power	182
	5.2	CapPlus, CapMinus	182
	5.3	ClockIn	182
	5.4	Reset	183
	5.5	Bootstrap	183
	5.6	Peek and poke	185
	5.7	Analyse	185
	5.8	Error	186
6		Memory	187
7		External memory interface	189
	7.1	ProcClockOut	189
	7.2	Tstates	190
	7.3	Internal access	190
	7.4	MemA0-15	190
	7.5	MemD0-15	190
	7.6	notMemWrB0-1	191
	7.7	notMemCE	193
	7.8	MemBAcc	195
	7.9	MemWait	196
	7.10	MemReq, MemGranted	198
8		Events	200
9		Links	201
10		Electrical specifications	204
	10.1	DC electrical characteristics	204
	10.2	Equivalent circuits	205
	10.3	AC timing characteristics	206
	10.4	Power rating	207
11		Package specifications	208
	11.1	68 pin grid array package	208
	11.2	68 pin PLCC J-bend package	210

6	**IMS M212 Preview**			213
	1		Introduction	214
		1.1	IMS M212 peripheral processor	215
			1.1.1 Central processor	215
			1.1.2 Peripheral interface	215
			1.1.3 Disk controller	215
			1.1.4 Links	216
			1.1.5 Memory system	216
			1.1.6 Error handling	216

	2	Operation		217
		2.1	Mode 1	217
		2.2	Mode 2	218
	3	Applications		219
	4	Package specifications		223
		4.1	68 pin grid array package	223
		4.2	68 pin PLCC J-bend package	225

7 IMS C004 Engineering Data   227

1	Introduction		228
2	Pin designations		230
3	System services		231
	3.1	Power	231
	3.2	CapPlus, CapMinus	231
	3.3	ClockIn	231
	3.4	Reset	233
4	Links		234
5	Switch implementation		237
6	Applications		238
	6.1	Link switching	238
	6.2	Multiple IMS C004 control	238
	6.3	Bidirectional exchange	238
	6.4	Bus systems	238
7	Electrical specifications		242
	7.1	DC electrical characteristics	242
	7.2	Equivalent circuits	243
	7.3	AC timing characteristics	244
	7.4	Power rating	244
8	Package specifications		245
	8.1	84 pin grid array package	245
9	IMS C004-A		247

8 IMS C011 Engineering Data   249

1	Introduction	250
2	Pin designations	252

	3	System services	253
	3.1	Power	253
	3.2	CapMinus	253
	3.3	ClockIn	253
	3.4	SeparateIQ	254
	3.5	Reset	255
	4	Links	256
	5	Mode 1 parallel interface	259
	5.1	Input port	259
	5.2	Output port	260
	6	Mode 2 parallel interface	261
	6.1	D0-7	261
	6.2	notCS	261
	6.3	RnotW	261
	6.4	RS0-1	261
		6.4.1 Input Data Register	261
		6.4.2 Input Status Register	264
	6.5	InputInt	264
		6.5.1 Output Data Register	264
		6.5.2 Output Status Register	264
	6.6	OutputInt	265
	6.7	Data read	265
	6.8	Data write	265
	7	Electrical specifications	266
	7.1	DC electrical characteristics	266
	7.2	Equivalent circuits	267
	7.3	AC timing characteristics	268
	7.4	Power rating	269
	8	Package specifications	270
	8.1	28 pin plastic dual-in-line package	270
	8.2	28 pin ceramic dual-in-line package	271
	8.3	Pinout	272

9 IMS C012 Engineering Data 273

	1	Introduction	274
	2	Pin designations	276
	3	System services	277
	3.1	Power	277
	3.2	CapMinus	277
	3.3	ClockIn	277
	3.4	Reset	279
	4	Links	280

	5	Parallel interface		283
		5.1	D0-7	283
		5.2	notCS	283
		5.3	RnotW	283
		5.4	RS0-1	283
			5.4.1 Input Data Register	283
			5.4.2 Input Status Register	286
		5.5	InputInt	286
			5.5.1 Output Data Register	286
			5.5.2 Output Status Register	286
		5.6	OutputInt	287
		5.7	Data read	287
		5.8	Data write	287
	6	Electrical specifications		288
		6.1	DC electrical characteristics	288
		6.2	Equivalent circuits	289
		6.3	AC timing characteristics	290
		6.4	Power rating	291
	7	Package specifications		292
		7.1	24 pin plastic dual-in-line package	292
		7.2	Pinout	293

Appendices

	A	Performance		295
		A.1	Performance overview	297
		A.2	Fast multiply, TIMES	299
		A.3	Arithmetic	300
		A.4	IMS T212, IMS T414 floating point operations	300
		A.5	IMS T800 floating point operations	301
			A.5.1 IMS T800 floating point functions	302
			A.5.2 IMS T800 special purpose functions and procedures	302
		A.6	Effect of external memory	303
		A.7	Interrupt latency	304
	B	Instruction Set Summary		305
	C	Bibliography		315
		C.1	INMOS publications	317
		C.2	INMOS technical notes	318
		C.3	Papers and extracts by INMOS authors	319
		C.4	Papers and extracts by other authors	320
		C.5	Books and monographs	323
		C.6	References	324
	D	Index		325
	E	Ordering		343

Preface

This reference manual describes the architecture of the transputer family of products and details some of the devices which make up that family. Items described include the 32 bit and 16 bit transputer products IMS T800, IMS T414 and IMS T212; the peripheral controller IMS M212; and the communications devices IMS C004, IMS C011 and IMS C012.

The manual first describes the transputer architecture and general features of transputer family devices. It then continues with the various product data sheets, followed by comparative performance details.

A transputer is a single VLSI device with processor, memory and communications links for direct connection to other transputers. Concurrent systems can be constructed from a collection of transputers operating concurrently and communicating through links. The transputer can be used as a building block for concurrent processing systems, with occam as the associated design formalism.

Current transputer products include the 16 bit IMS T212, the 32 bit IMS T414 and the IMS T800, a 32 bit transputer similar to the IMS T414 but with an integral high speed floating point processor.

The IMS M212 is an intelligent peripheral controller. It contains a 16 bit processor, on-chip memory and communications links. It contains hardware and interface logic to control disk drives and can be used as a programmable disk controller or as a general purpose peripheral interface.

The INMOS serial communication link is a high speed system interconnect which provides full duplex communication between members of the transputer family. It can also be used as a general purpose interconnect even where transputers are not used. The IMS C011 and IMS C012 link adaptors are communications devices enabling the INMOS serial communication link to be connected to parallel data ports and microprocessor buses. Ths IMS C004 is a programmable link switch. It provides a full crossbar switch between 32 link inputs and 32 link outputs.

The transputer development system referred to in this manual comprises an integrated editor, compiler and debugging system which enables transputers to be programmed in occam and in industry standard languages. The *Transputer Development System Manual* is supplied with the transputer development system.

Other information relevant to all transputer products is contained in the occam *Reference Manual*, supplied with INMOS software products and available as a separate publication. Where access to transputers at machine instruction set level is necessary, the document *The Transputer Instruction Set - A Compiler Writers' Guide* is available.

Various application and technical notes are also available from INMOS.

Software and hardware examples given in this manual are outline design studies and are included to illustrate various ways in which transputers can be used. The examples are not intended to provide accurate application designs.

Notation and nomenclature

The nomenclature and notation in general use throughout this manual is described below.

Significance

The bits in a byte are numbered 0 to 7, with bit 0 least significant. The bytes in words are numbered from 0, with byte 0 least significant. In general, wherever a value is treated as a number of component values, the components are numbered in order of increasing numerical significance, with the least significant component numbered 0. Where values are stored in memory, the least significant component value is stored at the lowest (most negative) address. Similarly, components of arrays are numbered starting from 0 and stored in memory with component 0 at the lowest address.

Transputer memory is byte addressed, with words aligned on four-byte boundaries for 32 bit devices and on two-byte boundaries for 16 bit devices.

Hexadecimal values are prefixed with #, as in *#1DF*.

Where a byte is transmitted serially, it is always transmitted least significant bit (0) first. In general, wherever a value is transmitted as a number of component values, the least significant component is transmitted first. Where an array is transmitted serially, component 0 is transmitted first. Consequently, block transfers to and from memory are performed starting with the lowest (most negative) address and ending with the highest (most positive) one.

In diagrams, the least significant component of a value is to the right hand side of the diagram. Component 0 of an array is at the bottom of a diagram, as are the most negative memory locations.

Signal naming conventions

Signal names identifying individual pins of a transputer chip have been chosen to avoid being cryptic, giving as much information as possible. The majority of transputer signals are active high. Those which are active low have names commencing with **not**; names such as **RnotW** imply that the first component of the name refers to its active high state and the second to its active low state. Capitals are used to introduce new components of a name, as in **ProcClockOut**.

All transputer signals described in the text of this manual are printed in **bold**. Registers and flags internal to a device are printed in *italics*, as are instruction operation codes. *Italics* are also used for emphasis. occam program notation is printed in a `fixed space teletype` style.

References

The manual is divided into several *chapters*, each chapter having a number of *sections* and *subsections*. Figures and tables have reference numbers tied to relevant sections of a particular chapter of the manual. Unless otherwise stated, all references refer to those within the current chapter of the manual.

Transputer product numbers

All INMOS products, both memories and transputers, have a part number of the general form

IMS abbbc-xyyz

Field **a** identifies the product group. This is a digit for memory products and a letter for other devices, the particular letter indicating the type of product (table .1). Field **bbb** identifies the product within that group and field **c** is its revision code. Field **x** denotes the package type, whilst field **yy** indicates speed variants etc. The final field **z** indicates to which specification the component is qualified; standard, military etc. Where appropriate some identifiers may be omitted, depending on the device.

A typical product part would be   IMS T414B-G15S.

Table .1: Transputer products

IMS 1...	Static RAM products
IMS A...	Digital signal processors
IMS B...	PC boards and modular hardware
IMS C...	Communications adaptors
IMS D...	Development system
IMS G...	Graphics products
IMS L...	Literature
IMS M...	Peripheral control transputers
IMS P...	occam programming system
IMS S...	Software product
IMS T...	Transputers

Chapter 1

transputer architecture

1 Introduction

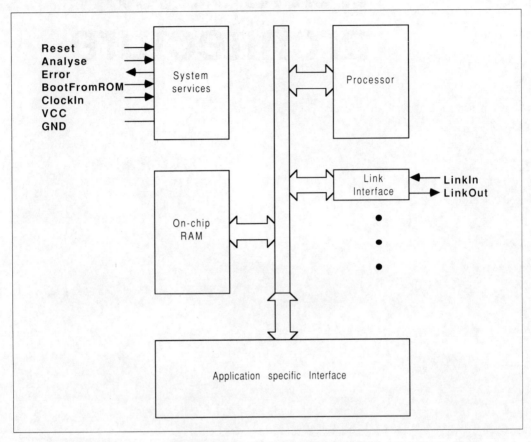

Figure 1.1: Transputer architecture

1 Introduction

1.1 Overview

A transputer is a microcomputer with its own local memory and with links for connecting one transputer to another transputer.

The transputer architecture defines a family of programmable VLSI components. The definition of the architecture falls naturally into the *logical* aspects which define how a system of interconnected transputers is designed and programmed, and the *physical* aspects which define how transputers, as VLSI components, are interconnected and controlled.

A typical member of the transputer product family is a single chip containing processor, memory, and communication links which provide point to point connection between transputers. In addition, each transputer product contains special circuitry and interfaces adapting it to a particular use. For example, a peripheral control transputer, such as a graphics or disk controller, has interfaces tailored to the requirements of a specific device.

A transputer can be used in a single processor system or in networks to build high performance concurrent systems. A network of transputers and peripheral controllers is easily constructed using point-to-point communication.

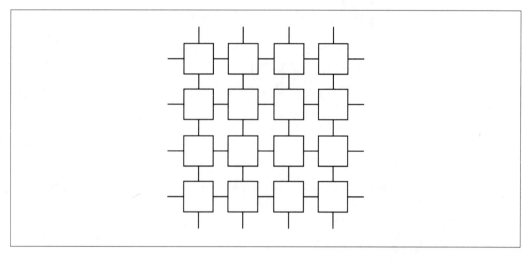

Figure 1.2: Transputer network

Transputers and occam

Transputers can be programmed in most high level languages, and are designed to ensure that compiled programs will be efficient. Where it is required to exploit concurrency, but still to use standard languages, occam can be used as a harness to link modules written in the selected languages.

To gain most benefit from the transputer architecture, the whole system can be programmed in occam (pages 8, 26). This provides all the advantages of a high level language, the maximum program efficiency and the ability to use the special features of the transputer.

occam provides a framework for designing concurrent systems using transputers in just the same way that boolean algebra provides a framework for designing electronic systems from logic gates. The system designer's task is eased because of the architectural relationship between occam and the transputer. A program running in a transputer is formally equivalent to an occam process, so that a network of transputers can be described directly as an occam program.

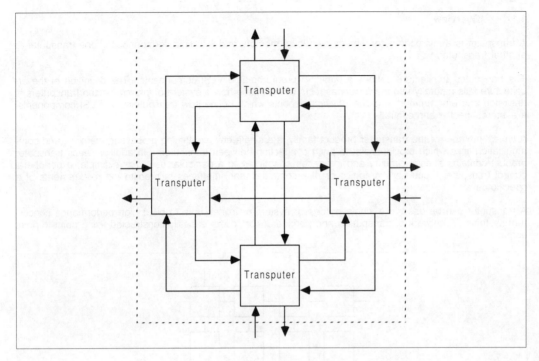

Figure 1.3: A node of four transputers

1.2 System design rationale

The transputer architecture simplifies system design by the use of processes as standard software and hardware building blocks.

An entire system can be designed and programmed in occam, from system configuration down to low level I/O and real time interrupts.

1.2.1 Programming

The software building block is the process. A system is designed in terms of an interconnected set of processes. Each process can be regarded as an independent unit of design. It communicates with other processes along point-to-point channels. Its internal design is hidden, and it is completely specified by the messages it sends and receives. Communication between processes is synchronized, removing the need for any separate synchronisation mechanism.

Internally, each process can be designed as a set of communicating processes. The system design is therefore hierarchically structured. At any level of design, the designer is concerned only with a small and manageable set of processes.

occam is based on these concepts, and provides the definition of the transputer architecture from the logical point of view (pages 8, 26).

1 Introduction

1.2.2 Hardware

Processes can be implemented in hardware. A transputer, executing an occam program, is a hardware process. The process can be independently designed and compiled. Its internal structure is hidden and it communicates and synchronizes with other transputers via its links, which implement occam channels.

Other hardware implementations of the process are possible. For example, a transputer with a different instruction set may be used to provide a different cost/performance trade-off. Alternatively, an implementation of the process may be designed in terms of hard-wired logic for enhanced performance.

The ability to specify a hard-wired function as an occam process provides the architectural framework for transputers with specialized capabilities (e.g., graphics). The required function (e.g., a graphics drawing and display engine) is defined as an occam process, and implemented in hardware with a standard occam channel interface. It can be simulated by an occam implementation, which in turn can be used to test the application on a development system.

1.2.3 Programmable components

A transputer can be programmed to perform a specialized function, and be regarded as a 'black box' thereafter. Some processes can be hard-wired for enhanced performance.

A system, perhaps constructed on a single chip, can be built from a combination of software processes, pre-programmed transputers and hardware processes. Such a system can, itself, be regarded as a component in a larger system.

The architecture has been designed to permit a network of programmable components to have any desired topology, limited only by the number of links on each transputer. The architecture minimizes the constraints on the size of such a system, and the hierarchical structuring provided by occam simplifies the task of system design and programming.

The result is to provide new orders of magnitude of performance for any given application, which can now exploit the concurrency provided by a large number of programmable components.

1.3 Systems architecture rationale

1.3.1 Point to point communication links

The transputer architecture simplifies system design by using point to point communication links. Every member of the transputer family has one or more standard links, each of which can be connected to a link of some other component. This allows transputer networks of arbitrary size and topology to be constructed.

Point to point communication links have many advantages over multi-processor buses:

> There is no contention for the communication mechanism, regardless of the number of transputers in the system.
>
> There is no capacitive load penalty as transputers are added to a system.
>
> The communications bandwidth does not saturate as the size of the system increases. Rather, the larger the number of transputers in the system, the higher the total communications bandwidth of the system. However large the system, all the connections between transputers can be short and local.

1.3.2 Local memory

Each transputer in a system uses its own local memory. Overall memory bandwidth is proportional to the number of transputers in the system, in contrast to a large global memory, where the additional processors must share the memory bandwidth.

Because memory interfaces are not shared, and are separate from the communications interfaces, they can be individually optimized on different transputer products to provide high bandwidth with the minimum of external components.

1.4 Communication

To provide synchronised communication, each message must be acknowledged. Consequently, a link requires at least one signal wire in each direction.

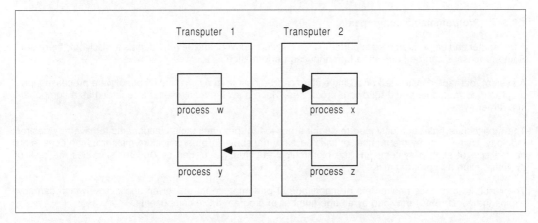

Figure 1.4: Links communicating between processes

A link between two transputers is implemented by connecting a link interface on one transputer to a link interface on the other transputer by two one-directional signal lines, along which data is transmitted serially.

The two signal wires of the link can be used to provide two occam channels, one in each direction. This requires a simple protocol. Each signal line carries data and control information.

The link protocol provides the synchronized communication of occam. The use of a protocol providing for the transmission of an arbitrary sequence of bytes allows transputers of different word length to be connected.

Each message is transmitted as a sequence of single byte communications, requiring only the presence of a single byte buffer in the receiving transputer to ensure that no information is lost. Each byte is transmitted as a start bit followed by a one bit followed by the eight data bits followed by a stop bit. After transmitting a data byte, the sender waits until an acknowledge is received; this consists of a start bit followed by a zero bit. The acknowledge signifies both that a process was able to receive the acknowledged byte, and that the receiving link is able to receive another byte. The sending link reschedules the sending process only after the acknowledge for the final byte of the message has been received.

Data bytes and acknowledges are multiplexed down each signal line. An acknowledge can be transmitted as soon as reception of a data byte starts (if there is room to buffer another one). Consequently transmission may be continuous, with no delays between data bytes.

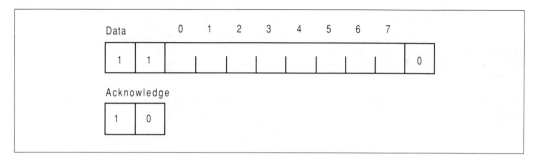

Figure 1.5: Link protocol

The links are designed to make the engineering of transputer systems straightforward. Board layout of two wire connections is easy to design and area efficient. All transputers will support a standard communications frequency of 10 Mbits/sec, regardless of processor performance. Thus transputers of different performance can be directly connected and future transputer systems will directly communicate with those of today.

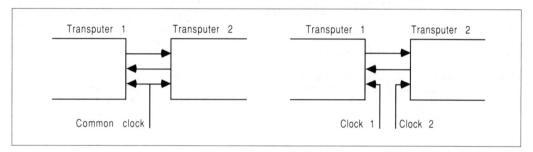

Figure 1.6: Clocking transputers

Link communication is not sensitive to clock phase. Thus, communication can be achieved between independently clocked systems as long as the communications frequency is the same.

The transputer family includes a number of link adaptor devices which provide a means of interfacing transputer links to non-transputer devices.

2   occam **model**

The programming model for transputers is defined by occam(page 26). The purpose of this section is to describe how to access and control the resources of transputers using occam. A more detailed description is available in the occam programming manual and the transputer development system manual (provided with the development system).

The transputer development system will enable transputers to be programmed in other industry standard languages. Where it is required to exploit concurrency, but still to use standard languages, occam can be used as a harness to link modules written in the selected languages.

2.1   Overview

In occam processes are connected to form concurrent systems. Each process can be regarded as a black box with internal state, which can communicate with other processes using point to point communication channels. Processes can be used to represent the behaviour of many things, for example, a logic gate, a microprocessor, a machine tool or an office.

The processes themselves are finite. Each process starts, performs a number of actions and then terminates. An action may be a set of sequential processes performed one after another, as in a conventional programming language, or a set of parallel processes to be performed at the same time as one another. Since a process is itself composed of processes, some of which may be executed in parallel, a process may contain any amount of internal concurrency, and this may change with time as processes start and terminate.

Ultimately, all processes are constructed from three primitive processes - assignment, input and output. An assignment computes the value of an expression and sets a variable to the value. Input and output are used for communicating between processes. A pair of concurrent processes communicate using a one way channel connecting the two processes. One process outputs a message to the channel and the other process inputs the message from the channel.

The key concept is that communication is synchronized and unbuffered. If a channel is used for input in one process, and output in another, communication takes place when both processes are ready. The value to be output is copied from the outputting process to the inputting process, and the inputting and outputting processes then proceed. Thus communication between processes is like the handshake method of communication used in hardware systems.

Since a process may have internal concurrency, it may have many input channels and output channels performing communication at the same time.

Every transputer implements the occam concepts of concurrency and communication. As a result, occam can be used to program an individual transputer or to program a network of transputers. When occam is used to program an individual transputer, the transputer shares its time between the concurrent processes and channel communication is implemented by moving data within the memory. When occam is used to program a network of transputers, each transputer executes the process allocated to it. Communication between occam processes on different transputers is implemented directly by transputer links. Thus the same occam program can be implemented on a variety of transputer configurations, with one configuration optimized for cost, another for performance, or another for an appropriate balance of cost and performance.

The transputer and occam were designed together. All transputers include special instructions and hardware to provide maximum performance and optimal implementations of the occam model of concurrency and communications.

All transputer instruction sets are designed to enable simple, direct and efficient compilation of occam. Programming of I/O, interrupts and timing is standard on all transputers and conforms to the occam model.

Different transputer variants may have different instruction sets, depending on the desired balance of cost, performance, internal concurrency and special hardware. The occam level interface will, however, remain standard across all products.

2 occam model

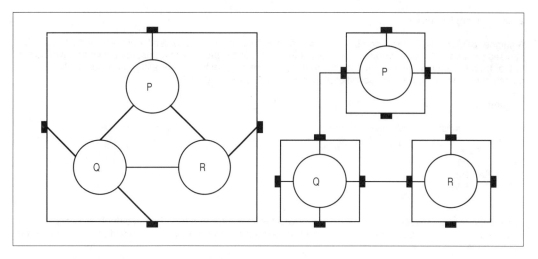

Figure 2.1: Mapping processes onto one or several transputers

2.2 occam overview

2.2.1 Processes

After it starts execution, a process performs a number of actions, and then either stops or terminates. Each action may be an assignment, an input, or an output. An assignment changes the value of a variable, an input receives a value from a channel, and an output sends a value to a channel.

At any time between its start and termination, a process may be ready to communicate on one or more of its channels. Each channel provides a one way connection between two concurrent processes; one of the processes may only output to the channel, and the other may only input from it.

Assignment

An assignment is indicated by the symbol :=. The example

    v := e

sets the value of the variable **v** to the value of the expression **e** and then terminates, for example: x := 0 sets **x** to zero, and x := x + 1 increases the value of **x** by 1.

Input

An input is indicated by the symbol ? The example

    c ? x

inputs a value from the channel **c**, assigns it to the variable **x** and then terminates.

Output

An output is indicated by the symbol ! The example

    c ! e

outputs the value of the expression **e** to the channel **c**.

2.2.2 Constructions

A number of processes can be combined to form a construct. A construct is itself a process and can therefore be used as a component of another construct. Each component process of a construct is written two spaces further from the left hand margin, to indicate that it is part of the construct. There are four classes of constructs namely the sequential, parallel, conditional and the alternative construct.

Sequence

A sequential construct is represented by

```
SEQ
  P1
  P2
  P3
  ...
```

The component processes P1, P2, P3 ... are executed one after another. Each component process starts after the previous one terminates and the construct terminates after the last component process terminates. For example

```
SEQ
  c1 ? x
  x  := x + 1
  c2 ! x
```

inputs a value, adds one to it, and then outputs the result.

Sequential constructs in occam are similar to programs written in conventional programming languages. Note, however, that they provide the performance and efficiency equivalent to that of an assembler for a conventional microprocessor.

Parallel

A parallel construct is represented by

```
PAR
  P1
  P2
  P3
  ...
```

The component processes P1, P2, P3 ... are executed together, and are called concurrent processes. The construct terminates after all of the component processes have terminated, for example:

```
PAR
  c1 ? x
  c2 ! y
```

allows the communications on channels c1 and c2 to take place together.

The parallel construct is unique to occam. It provides a straightforward way of writing programs which directly reflects the concurrency inherent in real systems. The implementation of parallelism on a single transputer is highly optimized so as to incur minimal process scheduling overhead.

2 occam model

Communication

Concurrent processes communicate only by using channels, and communication is synchronized. If a channel is used for input in one process, and output in another, communication takes place when both the inputting and the outputting processes are ready. The value to be output is copied from the outputting process to the inputting process, and the processes then proceed.

Communication between processes on a single transputer is via memory-to-memory data transfer. Between processes on different transputers it is via standard links. In either case the occam program is identical.

Conditional

A conditional construct

```
IF
  condition1
    P1
  condition2
    P2
  ...
```

means that **P1** is executed if **condition1** is true, otherwise **P2** is executed if **condition2** is true, and so on. Only one of the processes is executed, and then the construct terminates, for example:

```
IF
  x = 0
    y := y + 1
  x <> 0
    SKIP
```

increases **y** only if the value of **x** is 0.

Alternation

An alternative construct

```
ALT
  input1
    P1
  input2
    P2
  input3
    P3
  ...
```

waits until one of **input1**, **input2**, **input3** ... is ready. If **input1** first becomes ready, **input1** is performed, and then process **P1** is executed. Similarly, if **input2** first becomes ready, **input2** is performed, and then process **P2** is executed. Only one of the inputs is performed, then its corresponding process is executed and then the construct terminates, for example:

```
ALT
  count ? signal
    counter := counter + 1
  total ? signal
    SEQ
      out ! counter
      counter := 0
```

either inputs a signal from the channel **count**, and increases the variable **counter** by 1, or alternatively inputs from the channel **total**, outputs the current value of the counter, then resets it to zero.

The **ALT** construct provides a formal language method of handling external and internal events that must be handled by assembly level interrupt programming in conventional microprocessors.

Loop

```
WHILE condition
    P
```

repeatedly executes the process **P** until the value of the condition is false, for example:

```
WHILE (x - 5) > 0
    x := x - 5
```

leaves **x** holding the value of (**x** remainder 5) if **x** were positive.

Selection

A selection construct

```
CASE s
  n
    P1
  m, q
    P2
  ...
```

means that **P1** is executed if **s** has the same value as **n**, otherwise **P2** is executed if **s** has the same value as **m** or **q**, and so on, for example:

```
CASE direction
  up
    x := x + 1
  down
    x := x - 1
```

increases the value of **x** if **direction** is equal to **up**, otherwise if **direction** is equal to **down** the value of **x** is decreased.

Replication

A replicator is used with a **SEQ**, **PAR**, **IF** or **ALT** construction to replicate the component process a number of times. For example, a replicator can be used with **SEQ** to provide a conventional loop.

```
SEQ i = 0 FOR n
    P
```

causes the process **P** to be executed **n** times.

A replicator may be used with **PAR** to construct an array of concurrent processes.

```
PAR i = 0 FOR n
    Pi
```

constructs an array of **n** similar processes **P0**, **P1**, ..., **Pn-1**. The index **i** takes the values 0, 1, ..., n-1, in **P0**, **P1**, ..., **Pn-1** respectively.

2.2.3 Types

Every variable, expression and value has a type, which may be a primitive type, array type, record type or variant type. The type defines the length and interpretation of data.

All implementations provide the primitive types shown in table 2.1.

Table 2.1: Types

CHAN OF *protocol*	Each communication channel provides communication between two concurrent processes. Each channel is of a type which allows communication of data according to the specified protocol.
TIMER	Each timer provides a clock which can be used by any number of concurrent processes.
BOOL	The values of type **BOOL** are true and false.
BYTE	The values of type **BYTE** are unsigned numbers **n** in the range $0 <= n < 256$.
INT	Signed integers **n** in the range $-2^{31} <= n < 2^{31}$.
INT16	Signed integers **n** in the range $-2^{15} <= n < 2^{15}$.
INT32	Signed integers **n** in the range $-2^{31} <= n < 2^{31}$.
INT64	Signed integers **n** in the range $-2^{63} <= n < 2^{63}$.
REAL32	Floating point numbers stored using a sign bit, 8 bit exponent and 23 bit fraction in ANSI/IEEE Standard 754-1985 representation.
REAL64	Floating point numbers stored using a sign bit, 11 bit exponent and 52 bit fraction in ANSI/IEEE Standard 754-1985 representation.

2.2.4 Declarations, arrays and subscripts

A declaration **T x** declares **x** as a new channel, variable, timer or array of type **T**, for example:

    INT x:
    P

declares **x** as an integer variable for use in process **P**.

Array types are constructed from component types. For example [n] **T** is an array type constructed from **n** components of type **T**.

A component of an array may be selected by subscription, for example **v[e]** selects the **e**'th component of **v**.

A set of components of an array may be selected by subscription, for example [**v FROM e FOR c**] selects the **c** components **v[e], v[e + 1], ... v[e + c - 1]**. A set of components of an array may be assigned, input or output.

2.2.5 Procedures

A process may be given a name, for example:

```
PROC square (INT n)
  n := n * n
:
```

defines the procedure **square**. The name may be used as an instance of the process, for example:

```
square (x)
```

is equivalent to

```
n IS x:
n := n * n
```

2.2.6 Functions

A function can be defined in the same way as a procedure. For example:

```
INT FUNCTION factorial (VAL INT n)
  INT product:
  VALOF
    IF
      n >= 0
        SEQ
          product := 1
          SEQ i = 1 FOR n
            product := product * i
    RESULT product
:
```

defines the function **factorial**, which may appear in expressions such as

```
m := factorial (6)
```

2.2.7 Expressions

An expression is constructed from the operators given in table 2.2, from variables, numbers, the truth values **TRUE** and **FALSE**, and the brackets **(** and **)**.

Table 2.2: Operators

Operator	Operand types	Description
+ - * / REM	integer, real	arithmetic operators
PLUS MINUS TIMES AFTER	integer	modulo arithmetic
= <>	any primitive	relational operators
> < >= <=	integer, real	relational operators
AND OR NOT	boolean	boolean operators
/\ \/ >< ~	integers	bitwise operators: and, or, xor, not
<< >>	integer	shift operators

For example, the expression

    `(5 + 7) / 2`

evaluates to 6, and the expression

    `(#1DF /\ #F0) >> 4`

evaluates to `#D` (the character `#` introduces a hexadecimal constant).

A string is represented as a sequence of ASCII characters, enclosed in double quotation marks ". If the string has **n** characters, then it is an array of type `[n]BYTE`.

2.2.8 Timer

All transputers incorporate a timer. The implementation directly supports the occam model of time. Each process can have its own independent timer, which can be used for internal measurement or for real time scheduling.

A timer input sets a variable to a value of type `INT` representing the time. The value is derived from a clock, which changes at regular intervals, for example:

    `tim ? v`

sets the variable **v** to the current value of a free running clock, declared as the timer `tim`.

A delayed input takes the following form

    `tim ? AFTER e`

A delayed input is unable to proceed until the value of the timer satisfies (*timer* **AFTER** *e*). The comparison performed is a modulo comparison. This provides the effect that, starting at any point in the timer's cycle, the previous half cycle of the timer is considered as being before the current time, and the next half cycle is considered as being after the current time.

2.2.9 Peripheral access

The implementation of occam provides for peripheral access by extending the input and output primitives with a port input/output mechanism. A port is used like an occam channel, but has the effect of transferring information to and from a block of addresses associated with a peripheral.

Ports behave like occam channels in that only one process may input from a port, and only one process may output to a port. Thus ports provide a secure method of accessing external memory mapped status registers etc.

Note that there is no synchronization mechanism associated with port input and output. Any timing constraints which result from the use of asynchronous external hardware will have to be programmed explicitly. For example, a value read by a port input may depend upon the time at which the input was executed, and inputting at an invalid time would produce unusable data.

During applications development it is recommended that the peripheral is modelled by an occam process connected via channels.

2.3 Configuration

occam programs may be configured for execution on one or many transputers. The transputer development system provides the necessary tools for correctly distributing a program configured for many transputers.

Configuration does not affect the logical behaviour of a program (see section four, Program development). However, it does enable the program to be arranged to ensure that performance requirements are met.

PLACED PAR

A parallel construct may be configured for a network of transputers by using the **PLACED PAR** construct. Each component process (termed a placement) is executed by a separate transputer. The variables and timers used in a placement must be declared within each placement process.

PRI PAR

On any individual transputer, the outermost parallel construct may be configured to prioritize its components. Each process is executed at a separate priority. The first process has the highest priority, the last process has the lowest priority. Lower priority components may only proceed when all higher priority components are unable to proceed.

2.3.1 INMOS standard links

Each link provides one channel in each direction between two transputers.

A channel (which must already have been declared) is associated with a link by a channel association, for example:

```
PLACE Link0Input AT 4 :
```

3 Error handling

Errors in occam programs are either detected by the compiler or can be handled at runtime in one of three ways.

1 Cause the process to **STOP** allowing other processes to continue.

2 Cause the whole system to halt.

3 Have an arbitrary (undefined) effect.

The occam process **STOP** starts but never terminates. In method **1**, an errant process stops and in particular cannot communicate erroneous data to other processes. Other processes will continue to execute until they become dependent on data from the stopped process. It is therefore possible, for example, to write a process which uses a timeout to warn of a stopped process, or to construct a redundant system in which several processes performing the same task are used to enable the system to continue after one of them has failed.

Method **1** is the preferred method of executing a program.

Method **2** is useful for program development and can be used to bring transputers to an immediate halt, preventing execution of further instructions. The transputer **Error** output can be used to inform the transputer development system that such an error has occurred. No variable local to the process can be overwritten with erroneous data, facilitating analysis of the program and data which gave rise to the error.

Method **3** is useful only for optimising programs which are known to be correct!

When a system has stopped or halted as a result of an error, the state of all transputers in the system can be analysed using the transputer development system.

For languages other than occam, the transputer provides facilities for handling individual errors by software.

4 Program development

The development of programs for multiple processor systems can involve experimentation. In some cases, the most effective configuration is not always clear until a substantial amount of work has been done. For this reason, it is desirable that most of the design and programming can be completed before hardware construction is started.

4.1 Logical behaviour

An important property of occam in this context is that it provides a clear notion of 'logical behaviour'; this relates to those aspects of a program not affected by real time effects.

It is guaranteed that the logical behaviour of a program is not altered by the way in which the processes are mapped onto processors, or by the speed of processing and communication. Consequently a program ultimately intended for a network of transputers can be compiled, executed and tested on a single computer used for program development.

Even if the application uses only a single transputer, the program can be designed as a set of concurrent processes which could run on a number of transputers. This design style follows the best traditions of structured programming; the processes operate completely independently on their own variables except where they explicitly interact, via channels. The set of concurrent processes can run on a single transputer or, for a higher performance product, the processes can be partitioned amongst a number of transputers.

It is necessary to ensure, on the development system, that the logical behaviour satisfies the application requirements. The only ways in which one execution of a program can differ from another in functional terms result from dependencies upon input data and the selection of components of an **ALT**. Thus a simple method of ensuring that the application can be distributed to achieve any desired performance is to design the program to behave 'correctly' regardless of input data and **ALT** selection.

4.2 Performance measurement

Performance information is useful to gauge overall throughput of an application, and has to be considered carefully in applications with real time constraints.

Prior to running in the target environment, an occam program should be relatively mature, and indeed should be correct except for interactions which do not obey the occam synchronization rules. These are precisely the external interactions of the program where the world will not wait to communicate with an occam process which is not ready. Thus the set of interactions that need to be tested within the target environment are well identified.

Because, in occam, every program is a process, it is extremely easy to add monitor processes or simulation processes to represent parts of the real time environment, and then to simulate and monitor the anticipated real time interactions. The occam concept of time and its implementation in the transputer is important. Every process can have an independent timer enabling, for example, all the real time interactions to be modelled by separate processes and any time dependent features to be simulated.

4.3 Separate compilation of occam and other languages

A program portion which is separately compiled, and possibly written in a language other than occam, may be executed on a single transputer.

If the program is written in occam, then it takes the form of a single **PROC**, with only channel parameters. If the program is written in a language other than occam, then a run-time system is provided which provides input/output to occam channels.

4 Program development

Such separately compiled program portions are linked together by a framework of channels, termed a harness. The harness is written in occam. It includes all configuration information, and in particular specifies the transputer configuration in which the separately compiled program portion is executed.

Transputers are designed to allow efficient implementations of high level languages, such as C, Pascal and Fortran. Such languages will be available in addition to occam.

At runtime, a program written in such a language is treated as a single occam process. Facilities are provided in the implementations of these languages to allow such a program to communicate on occam channels. It can thus communicate with other such programs, or with programs written in occam. These programs may reside on the same transputer, in which case the channels are implemented in store, or may reside on different transputers, in which case the channels are implemented by transputer links.

It is therefore possible to implement occam processes in conventional high level languages, and arrange for them to communicate. It is possible for different parts of the same application to be implemented in different high level languages.

The standard input and output facilities provided within these languages are implemented by a well-defined protocol of communications on occam channels.

The development system provides facilities for management of separately compiled occam.

4.4 Memory map and placement

The low level memory model is of a signed address space.

Memory is byte addressed, the lowest addressed byte occupying the least significant byte position within the word.

The implementation of occam supports the allocation of the code and data areas of an occam process to specific areas of memory. Such a process must be a separately compiled **PROC**, and must not reference any variables and timers other than those declared within it.

5 Physical architecture

5.1 INMOS serial links

5.1.1 Overview

All transputers have several links. The link protocol and electrical characteristics form a standard for all INMOS transputer and peripheral products.

All transputers support a standard link communications frequency of 10 megabits per second. Some devices also support other data rates. Maintaining a standard communications frequency means that devices of mixed performance and type can intercommunicate easily.

Each link consists of two unidirectional signal wires carrying both data and control bits. The link signals are TTL compatible so that their range can be easily extended by inserting buffers.

The INMOS communication links provide for communication between devices on the same printed circuit board or between printed circuit boards via a back plane. They are intended to be used in electrically quiet environments in the same way as logic signals between TTL gates.

The number of links, and any communication speeds in addition to the standard speed of 10 Mbits/sec, are given in the **product data** for each product.

5.1.2 Link electrical specification

The quiescent state of the link signals is low, for a zero. The link input signals and output signals are standard TTL compatible signals.

For correct functioning of the links the specifications for maximum variation in clock frequency between two transputers joined by a link and maximum capacitive load must be met. Each transputer product also has specified the maximum permissible variation in delay in buffering, and minimum permissible edge gradients. Details of these specifications are provided in the product data.

Provided that these specifications are met then any buffering employed may introduce an arbitrary delay into a link signal without affecting its correct operation.

5.2 System services

5.2.1 Powering up and down, running and stopping

At all times the specification of input voltages with respect to the **GND** and **VCC** pins must be met. This includes the times when the **VCC** pins are ramping to 5 V, and also while they are ramping from 5 V down to 0 V.

The system services comprise the clocks, power, and signals used for initialization.

The specification includes minimum times that **VCC** must be within specification, the input clock must be oscillating, and the **Reset** signal must be high before **Reset** goes low. These specifications ensure that internal clocks and logic have settled before the transputer starts.

When the transputer is reset the memory interface is initialised (if present and configurable).

The processor and INMOS serial links start after reset. The transputer obeys a bootstrap program which can either be in off-chip ROM or can be received from one of the links. How to specify where the bootstrap program is taken from depends upon the type of transputer being used. The program will normally load up a larger program either from ROM or from a peripheral such as a disk.

During power down, as during power up, the input and output pins must remain within specification with respect to both **GND** and **VCC**.

5 Physical architecture

A software error, such as arithmetic overflow, array bounds violation or divide by zero, causes an error flag to be set in the transputer processor. The flag is directly connected to the **Error** pin. Both the flag and the pin can be ignored, or the transputer stopped. Stopping the transputer on an error means that the error cannot cause further corruption.

As well as containing the error in this way it is possible to determine the state of the transputer and its memory at the time the error occurred.

5.2.2 Clock distribution

All transputers operate from a standard 5MHz input clock. High speed clocks are derived internally from the low frequency input to avoid the problems of distributing high frequency clocks. Within limits the mark-to-space ratio, the voltage levels and the transition times are immaterial. The limits on these are given in the product data for each product. The asynchronous data reception of the links means that differences in the clock phase between chips is unimportant.

The important characteristic of the transputer's input clock is its stability, such as is provided by a crystal oscillator. An R-C oscillator is inadequate. The edges of the clock should be monotonic (without kinks), and should not undershoot below -0.5 V.

5.3 Bootstrapping from ROM or from a link

The program which is executed after reset can either reside in ROM in the transputer's address space or it can be loaded via any one of the transputer's INMOS serial links.

The transputer bootstraps from ROM by transferring control to the top two bytes in memory, which will invariably contain a backward jump into ROM.

If bootstrapping from a link, the transputer bootstraps from the first link to receive a message. The first byte of the message is the count of the number of bytes of program which follow. The program is loaded into memory starting at a product dependent location *MemStart*, and then control is transferred to this address.

Messages subsequently arriving on other links are not acknowledged until the transputer processor obeys a process which inputs from them. The loading of a network of transputers is controlled by the transputer development system, which ensures that the first message each transputer receives is the bootstrap program.

5.4 Peripheral interfacing

All transputers contain one or more INMOS serial links. Certain transputer products also have other application specific interfaces. The peripheral control transputers contain specialized interfaces to control a specific peripheral or peripheral family.

In general, a transputer based application will comprise a number of transputers which communicate using INMOS links. There are three methods of communicating with peripherals.

The first is by employing peripheral control transputers (eg for graphics or disks), in which the transputer chip connects directly to the peripheral concerned (figure 5.1). The interface to the peripheral is implemented by special purpose hardware within the transputer. The application software in the transputer is implemented as an occam process, and controls the interface via occam channels linking the processor to the special purpose hardware.

The second method is by employing link adaptors (figure 5.2). These devices convert between a link and a specialized interface. The link adaptor is connected to the link of an appropriate transputer, which contains the application designer's peripheral device handler implemented as an occam process.

The third method is by memory mapping the peripheral onto the memory bus of a transputer (figure 5.3). The peripheral is controlled by memory accesses issued as a result of **PORT** inputs and outputs. The application designer's peripheral device handler provides a standard occam channel interface to the rest of the application.

The first transputers implement an event pin which provides a simple means for an external peripheral to request attention from a transputer.

In all three methods, the peripheral driver interfaces to the rest of the application via occam channels. Consequently, a peripheral device can be simulated by an occam process. This enables testing of all aspects of a transputer system before the construction of hardware.

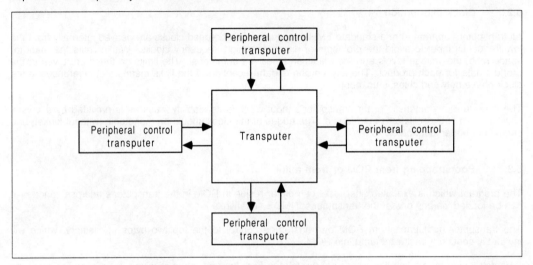

Figure 5.1: Transputer with peripheral control transputers

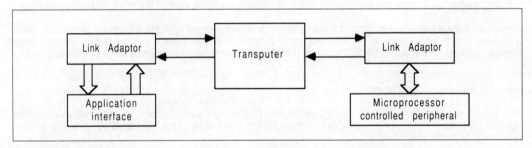

Figure 5.2: Transputer with link adaptors

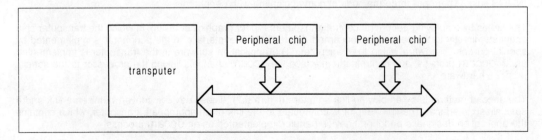

Figure 5.3: Memory mapped peripherals

Chapter 2

transputer overview

1   Introduction

The INMOS transputer family is a range of system components each of which combines processing, memory and interconnect in a single VLSI chip. A concurrent system can be constructed from a collection of transputers which operate concurrently and communicate through serial communication links. Such systems can be designed and programmed in occam, a language based on communicating processes. Transputers have been sucessfully used in application areas ranging from embedded systems to supercomputers.

The first member of the family, the IMS T414 32-bit transputer (Bibliography reference *INMOS '84*), was introduced in September 1985, and has enabled concurrency to be applied in a wide variety of applications such as simulation, robot control, image synthesis, and digital signal processing. Many computationally intensive applications can exploit large arrays of transputers; the system performance depending on the nunmber of transputers, the speed of inter-transputer communication and the performance of each transputer processor.

Many important applications of transputers involve floating point arithmetic. The latest addition to the INMOS transputer family, the IMS T800, can increase the performance of such systems by offering greatly improved floating-point and communication performance. The IMS T800-20, available in the second half of 1987, is capable of sustaining over one and a half million floating point operations per second; the IMS T800-30, available in the first half of 1988, is capable of sustaining over two and a quarter million floating point operations per second. The comparative figure for the IMS T414 transputer is somewhat less than one hundred thousand floating point operations per second.

For publication references used in this chapter, see page 324.

2 The transputer: basic architecture and concepts

2.1 A programmable device

The transputer is a component designed to exploit the potential of VLSI. This technology allows large numbers of *identical* devices to be manufactured cheaply. For this reason, it is attractive to implement a concurrent system using a number of identical components, each of which is customised by an appropriate program. The transputer is, therefore, a VLSI device with a processor, memory to store the program executed by the processor, and communication links for direct connection to other transputers. Transputer systems can be designed and programmed using occam which allows an application to be described as a collection of processes which operate concurrently and communicate through channels. The transputer can therefore be used as a building block for concurrent processing systems, with occam as the associated design formalism.

2.2 occam

occam enables a system to be described as a collection of concurrent processes, which communicate with each other and with peripheral devices through channels. occam programs are built from three primitive processes:

```
v := e        assign expression e to variable v
c ! e         output expression e to channel c
c ? v         input from channel c to variable v
```

The primitive processes are combined to form constructs:

```
SEQuential    components executed one after another
PARallel      components executed together
ALTernative   component first ready is executed
```

A construct is itself a process, and may be used as a component of another construct.

Conventional sequential programs can be expressed with variables and assignments, combined in sequential constructs. **IF** and **WHILE** constructs are also provided.

Concurrent programs can be expressed with channels, inputs and outputs, which are combined in parallel and alternative constructs.

Each occam channel provides a communication path between two concurrent processes. Communication is synchronised and takes place when both the inputting process and the outputting process are ready. The data to be output is then copied from the outputting process to the inputting process, and both processes continue.

An alternative process may be ready for input from any one of a number of channels. In this case, the input is taken from the channel which is first used for output by another process.

2.3 VLSI technology

One important property of VLSI technology is that communication between devices is very much slower than communication within a device. In a computer, almost every operation that the processor performs involves the use of memory. For this reason a transputer includes both processor and memory in the same integrated circuit device.

2 The transputer: basic architecture and concepts

In any system constructed from integrated circuit devices, much of the physical bulk arises from connections between devices. The size of the package for an integrated circuit is determined more by the number of connection pins than by the size of the device itself. In addition, connections between devices provided by paths on a circuit board consume a considerable amount of space.

The speed of communication between electronic devices is optimised by the use of one-directional signal wires, each connecting two devices. If many devices are connected by a shared bus, electrical problems of driving the bus require that the speed is reduced. Also, additional control logic and wiring are required to control sharing of the bus.

To provide maximum speed with minimal wiring, the transputer uses point-to-point serial communication links for direct connection to other transputers. The protocols used on the transputer links are discussed later.

2.4 Simplified processor with micro-coded scheduler

The most effective implementation of simple programs by a programmable computer is provided by a sequential processor. Consequently, the transputer has a fairly conventional microcoded processor. There is a small core of about 32 instructions which are used to implement simple sequential programs. In addition there are other, more specialised groups of instructions which provide facilities such as long arithmetic and process scheduling.

As a process executed by a transputer may itself consist of a number of concurrent processes the transputer has to support the occam programming model internally. The transputer, therefore, has a microcoded scheduler which shares the processor time between the concurrent processes. The scheduler provides two priority levels; any high priority process which can run will do so in preference to any low priority process.

2.5 Transputer products

The first transputer to become available was the INMOS IMS T414. This has a 32-bit processor, 2 Kbytes of fast on-chip memory, a 32-bit external memory interface and 4 links for connection to other transputers. The current fastest available version of this product, the IMS T414-20, has a 50 nS internal cycle time, and achieves about 10 MIPS on sequential programs. The second transputer to become available was the IMS T212; this is very similar to the IMS T414 but has a 16-bit processor and 16-bit external memory interface. The remaining transputer in the family is the IMS M212 disk processor. This contains a 16-bit processor, RAM, ROM, 2 inter-transputer links and special hardware to control both winchester and floppy disks.

In addition the transputer family includes a number of transputer link related products. There are the 'link adaptors' which convert between handshaken 8-bit parallel data and INMOS link bit-serial data. These allow transputers to be connected to conventional, bus-based systems, and also allow conventional microprocessors to use transputer links as a system interconnect. In addition there is the IMS C004, which is a link exchange.

3 Transputer internal architecture

Internally, the IMS T414 consists of a memory, processor and communications system connected via a 32-bit bus. The bus also connects to the external memory interface, enabling additional local memory to be used. The processor, memory and communications system each occupy about 25% of the total silicon area, the remainder being used for power distribution, clock generators and external connections.

The IMS T800, with its on-chip floating point unit, is only 20% larger in area than the IMS T414. The small size and high performance come from a design which takes careful note of silicon economics. This contrasts starkly with conventional co-processors, where the floating point unit typically occupies more area than a complete micro-processor, and requires a second chip.

The block diagram 3.1 indicates the way in which the major blocks of the IMS T800 and IMS T414 are interconnected.

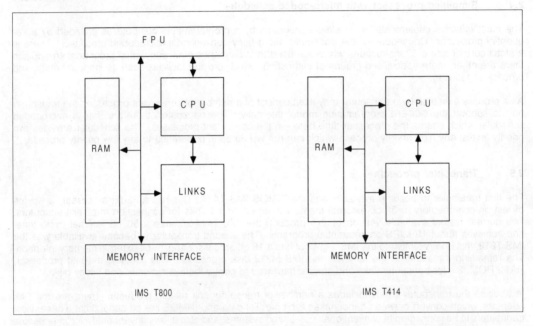

Figure 3.1: Transputer interconnections

The CPU of the transputers contains three registers (A, B and C) used for integer and address arithmetic, which form a hardware stack. Loading a value into the stack pushes B into C, and A into B, before loading A. Storing a value from A pops B into A and C into B. Similarly, the FPU includes a three register floating-point evaluation stack, containing the AF, BF, and CF registers. When values are loaded onto, or stored from the stack the AF, BF and CF registers push and pop in the same way as the A, B and C registers.

The addresses of floating point values are formed on the CPU stack, and values are transferred between the addressed memory locations and the FPU stack under the control of the CPU. As the CPU stack is used only to hold the addresses of floating point values, the wordlength of the CPU is independent of that of the FPU. Consequently, it would be possible to use the same FPU together with, for example, a 16-bit CPU such as that used on the IMS T212 transputer.

The transputer scheduler provides two priority levels. The FPU register stack is duplicated so that when the IMS T800 switches from low to high priority none of the state in the floating point unit is written to memory. This results in a worst-case interrupt response of only 2.5 μS (-30), or 3.7 μS (-20). Furthermore, the duplication of the register stack enables floating point arithmetic to be used in an interrupt routine without any performance penalty.

3 Transputer internal architecture

3.1 Sequential processing

The design of the transputer processor exploits the availability of fast on-chip memory by having only a small number of registers; the CPU contains six registers which are used in the execution of a sequential process. The small number of registers, together with the simplicity of the instruction set enables the processor to have relatively simple (and fast) data-paths and control logic.

The six registers are:

> The workspace pointer which points to an area of store where local variables are kept.
>
> The instruction pointer which points to the next instruction to be executed.
>
> The operand register which is used in the formation of instruction operands.

The A, B and C registers which form an evaluation stack, and are the sources and destinations for most arithmetic and logical operations. Loading a value into the stack pushes B into C, and A into B, before loading A. Storing a value from A, pops B into A and C into B.

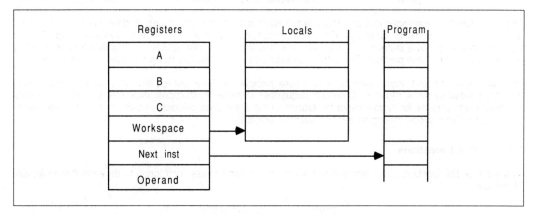

Figure 3.2: Registers

Expressions are evaluated on the evaluation stack, and instructions refer to the stack implicitly. For example, the *add* instruction adds the top two values in the stack and places the result on the top of the stack. The use of a stack removes the need for instructions to respecify the location of their operands. Statistics gathered from a large number of programs show that three registers provide an effective balance between code compactness and implementation complexity.

No hardware mechanism is provided to detect that more than three values have been loaded onto the stack. It is easy for the compiler to ensure that this never happens.

3.2 Instructions

It was a design decision that the transputer should be programmed in a high-level language. The instruction set has, therefore, been designed for simple and efficient compilation. It contains a relatively small number of instructions, all with the same format, chosen to give a compact representation of the operations most frequently occuring in programs. The instruction set is independant of the processor wordlength, allowing the same microcode to be used for transputers with different wordlengths. Each instruction consists of a single byte divided into two 4-bit parts. The four most significant bits of the byte are a function code, and the four least significant bits are a data value.

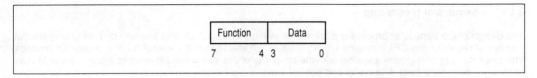

Figure 3.3: Instruction format

3.2.1 Direct functions

The representation provides for sixteen functions, each with a data value ranging from 0 to 15. Thirteen of these are used to encode the most important functions performed by any computer. These include:

load constant   add constant
load local   store local   load local pointer
load non-local   store non-local
jump   conditional jump   call

The most common operations in a program are the loading of small literal values, and the loading and storing of one of a small number of variables. The *load constant* instruction enables values between 0 and 15 to be loaded with a single byte instruction. The *load local* and *store local* instructions access locations in memory relative to the workspace pointer. The first 16 locations can be accessed using a single byte instruction.

The *load non-local* and *store non-local* instructions behave similarly, except that they access locations in memory relative to the A register. Compact sequences of these instructions allow efficient access to data structures, and provide for simple implementations of the static links or displays used in the implementation of block structured programming languages such as occam.

3.2.2 Prefix functions

Two more of the function codes are used to allow the operand of any instruction to be extended in length. These are:

prefix   negative prefix

All instructions are executed by loading the four data bits into the least significant four bits of the operand register, which is then used as the the instruction's operand. All instructions except the prefix instructions end by clearing the operand register, ready for the next instruction.

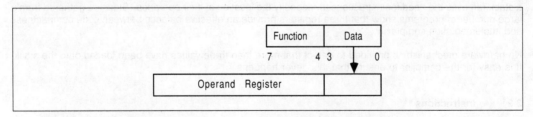

Figure 3.4: Instruction operand register

The *prefix* instruction loads its four data bits into the operand register, and then shifts the operand register up four places. The *negative prefix* instruction is similar, except that it complements the operand register before shifting it up. Consequently operands can be extended to any length up to the length of the operand register by a sequence of prefix instructions. In particular, operands in the range -256 to 255 can be represented using one prefix instruction.

3 Transputer internal architecture

The use of prefix instructions has certain beneficial consequences. Firstly, they are decoded and executed in the same way as every other instruction, which simplifies and speeds instruction decoding. Secondly, they simplify language compilation, by providing a completely uniform way of allowing any instruction to take an operand of any size. Thirdly, they allow operands to be represented in a form independent of the processor wordlength.

3.2.3 Indirect functions

The remaining function code, *operate*, causes its operand to be interpreted as an operation on the values held in the evaluation stack. This allows up to 16 such operations to be encoded in a single byte instruction. However, the prefix instructions can be used to extend the operand of an *operate* instruction just like any other. The instruction representation therefore provides for an indefinite number of operations.

The encoding of the indirect functions is chosen so that the most frequently occuring operations are represented without the use of a prefix instruction. These include arithmetic, logical and comparison operations such as

add   *exclusive or*   *greater than*

Less frequently occuring operations have encodings which require a single prefix operation (the transputer instruction set is not large enough to require more than 512 operations to be encoded!).

The IMS T800 has additional instructions which load into, operate on, and store from, the floating point register stack. It also contains new instructions which support colour graphics, pattern recognition and the implementation of error correcting codes. These instructions have been added whilst retaining the existing IMS T414 instruction set. This has been possible because of the extensible instruction encoding used in transputers.

3.2.4 Efficiency of encoding

Measurements show that about 70% of executed instructions are encoded in a single byte (ie without the use of prefix instructions). Many of these instructions, such as *load constant* and *add* require just one processor cycle.

The instruction representation gives a more compact representation of high level language programs than more conventional instruction sets. Since a program requires less store to represent it, less of the memory bandwidth is taken up with fetching instructions. Furthermore, as memory is word accessed the processor will receive several instructions for every fetch.

Short instructions also improve the effectiveness of instruction prefetch, which in turn improves processor performance. There is an extra word of prefetch buffer so that the processor rarely has to wait for an instruction fetch before proceeding. Since the buffer is short, there is little time penalty when a jump instruction causes the buffer contents to be discarded.

3.3 Support for concurrency

The processor provides efficient support for the occam model of concurrency and communication. It has a microcoded scheduler which enables any number of concurrent processes to be executed together, sharing the processor time. This removes the need for a software kernel. The processor does not need to support the dynamic allocation of storage as the occam compiler is able to perform the allocation of space to concurrent processes.

At any time, a concurrent process may be

    active   - being executed
                 - on a list waiting to be executed
    inactive - ready to input
                 - ready to output
                 - waiting until a specified time

The scheduler operates in such a way that inactive processes do not consume any processor time. The active processes waiting to be executed are held on a list. This is a linked list of process workspaces, implemented using two registers, one of which points to the first process on the list, the other to the last. In figure 3.5, S is executing, and P, Q and R are active, awaiting execution.

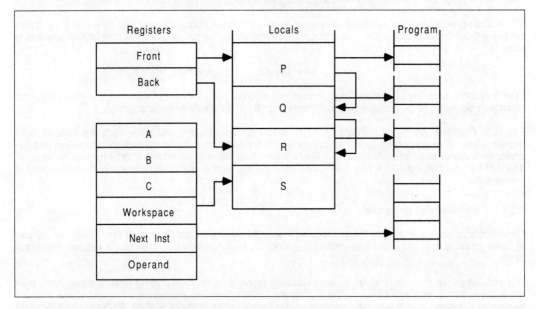

Figure 3.5: Linked process list

A process is executed until it is unable to proceed because it is waiting to input or output, or waiting for the timer. Whenever a process is unable to proceed, its instruction pointer is saved in its workspace and the next process is taken from the list. Actual process switch times are very small as little state needs to be saved; it is not necessary to save the evaluation stack on rescheduling.

The processor provides a number of special operations to support the process model. These include

start process   *end process*

When a parallel construct is executed, *start process* instructions are used to create the necessary concurrent processes. A *start process* instruction creates a new process by adding a new workspace to the end of the scheduling list, enabling the new concurrent process to be executed together with the ones already being executed.

The correct termination of a parallel construct is assured by use of the *end process* instruction. This uses a workspace location as a counter of the components of the parallel construct which have still to terminate. The counter is initialised to the number of components before the processes are 'started'. Each component ends with an *end process* instruction which decrements and tests the counter. For all but the last component, the counter is non zero and the component is descheduled. For the last component, the counter is zero and the component continues.

3 Transputer internal architecture

3.4 Communications

Communication between processes is achieved by means of channels. occam communication is point-to-point, synchronised and unbuffered. As a result, a channel needs no process queue, no message queue and no message buffer.

A channel between two processes executing on the same transputer is implemented by a single word in memory; a channel between processes executing on different transputers is implemented by point-to-point links. The processor provides a number of operations to support message passing, the most important being

input message   *output message*

The *input message* and *output message* instructions use the address of the channel to determine whether the channel is internal or external. This means that the same instruction sequence can be used for both hard and soft channels, allowing a process to be written and compiled without knowledge of where its channels are connected.

As in the occam model, communication takes place when both the inputting and outputting processes are ready. Consequently, the process which first becomes ready must wait until the second one is also ready.

A process performs an input or output by loading the evaluation stack with a pointer to a message, the address of a channel, and a count of the number of bytes to be transferred, and then executing an *input message* or an *output message* instruction.

3.4.1 Internal channel communication

At any time, an internal channel (a single word in memory) either holds the identity of a process, or holds the special value *empty*. The channel is initialised to *empty* before it is used.

When a message is passed using the channel, the identity of the first process to become ready is stored in the channel, and the processor starts to execute the next process from the scheduling list. When the second process to use the channel becomes ready, the message is copied, the waiting process is added to the scheduling list, and the channel reset to its initial state. It does not matter whether the inputting or outputting process becomes ready first.

In figure 3.6, a process P is about to execute an output instruction on an 'empty' channel C. The evaluation stack holds a pointer to a message, the address of channel C, and a count of the number of bytes in the message.

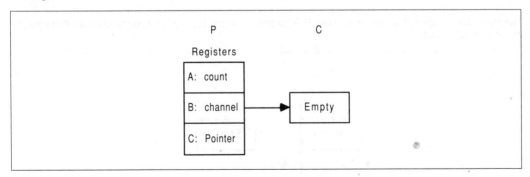

Figure 3.6: Output to empty channel

After executing the output instruction, the channel C holds the address of the workspace of P, and the address of the message to be transferred is stored in the workspace of P. P is descheduled, and the process starts to execute the next process from the scheduling list.

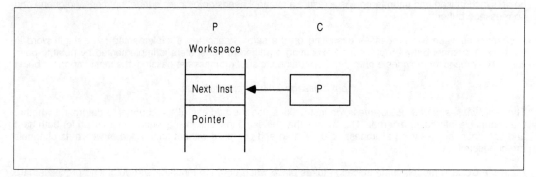

Figure 3.7:

The channel C and the process P remain in this state until a second process, Q executes an output instruction on the channel.

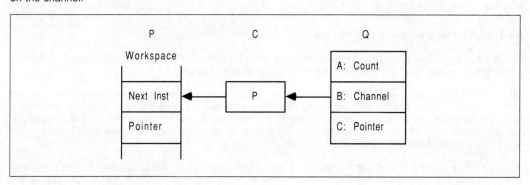

Figure 3.8:

The message is copied, the waiting process P is added to the scheduling list, and the channel C is reset to its initial 'empty' state.

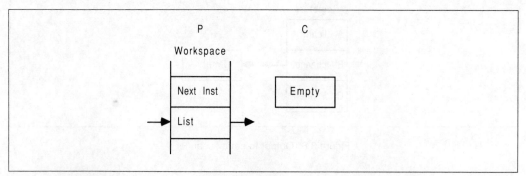

Figure 3.9:

3 Transputer internal architecture

3.4.2 External channel communication

When a message is passed via an external channel the processor delegates to an autonomous link interface the job of transferring the message and deschedules the process. When the message has been transferred the link interface causes the processor to reschedule the waiting process. This allows the processor to continue the execution of other processes whilst the external message transfer is taking place.

Each link interface uses three registers:

       a pointer to a process workspace
       a pointer to a message
       a count of bytes in the message

In figure 3.10 processes P and Q executed by different transputers communicate using a channel C implemented by a link connecting two transputers. P outputs, and Q inputs.

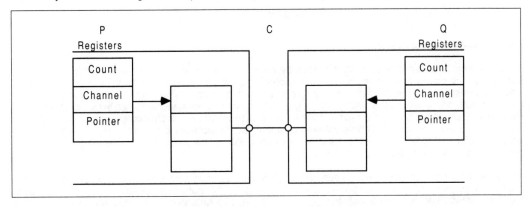

Figure 3.10: Communication between transputers

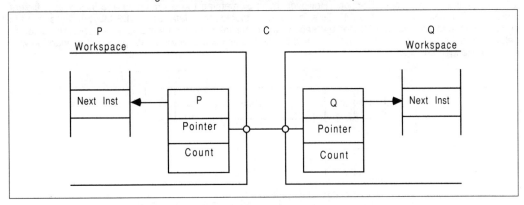

Figure 3.11:

When P executes its output instruction, the registers in the link interface of the transputer executing P are initialised, and P is descheduled. Similarly, when Q executes its input instruction, the registers in the link interface of the process executing Q are initialised, and Q is descheduled (figure 3.11).

The message is now copied through the link, after which the workspaces of P and Q are returned to the corresponding scheduling lists (figure 3.12). The protocol used on P and Q ensures that it does not matter which of P and Q first becomes ready.

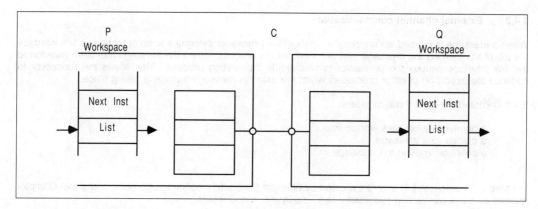

Figure 3.12:

3.4.3 Communication links

A link between two transputers is implemented by connecting a link interface on one transputer to a link interface on the other transputer by two one-directional signal wires, along which data is transmitted serially. The two wires provide two occam channels, one in each direction. This requires a simple protocol to multiplex data and control information. Messages are transmitted as a sequence of bytes, each of which must be acknowledged before the next is transmitted. A byte of data is transmitted as a start bit followed by a one bit followed by eight bits of data followed by a stop bit. An acknowledgement is transmitted as a start bit followed by a stop bit. An acknowledgement indicates both that a process was able to receive the data byte and that it is able to buffer another byte.

The protocol permits an acknowledgement to be generated as soon as the receiver has identified a data packet. In this way the acknowledgement can be received by the transmitter before all of the data packet has been transmitted and the transmitter can transmit the next data packet immediately. The IMS T414 transputer does not implement this overlapping and achieves a data rate of 0.8 Mbytes per second using a link to transfer data in one direction. However, by implementing the overlapping and including sufficient buffering in the link hardware, the IMS T800 more than doubles this data rate to 1.8 Mbytes per second in one direction, and achieves 2.4 Mbytes per second when the link carries data in both directions. The diagram below shows the signals that would be observed on the two link wires when a data packet is overlapped with an acknowledgement.

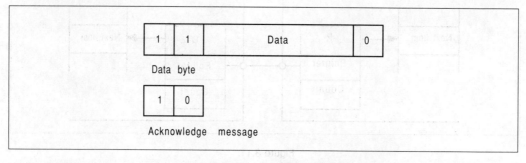

Figure 3.13: Link data and acknowledge formats

3   Transputer internal architecture

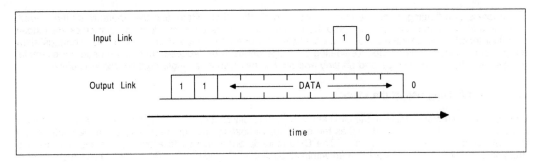

Figure 3.14: Overlapped link acknowledge

3.5   Timer

The transputer has a clock which 'ticks' every microsecond. The current value of the processor clock can be read by executing a *read timer* instruction.

A process can arrange to perform a *timer input*, in which case it will become ready to execute after a specified time has been reached.

The *timer input* instruction requires a time to be specified. If this time is in the 'past' (i.e. *ClockReg* **AFTER** *SpecifiedTime*) then the instruction has no effect. If the time is in the 'future' (i.e. *SpecifiedTime* **AFTER** *Clockreg* or *SpecifiedTime* = *ClockReg*) then the process is descheduled. When the specified time is reached the process is scheduled again.

3.6   Alternative

The occam alternative construct enables a process to wait for input from any one of a number of channels, or until a specific time occurs. This requires special instructions, as the normal *input* instruction deschedules a process until a specific channel becomes ready, or until a specific time is reached. The instructions are:

> enable channel   disable channel
> enable timer   disable timer
> alternative wait

The alternative is implemented by 'enabling' the channel input or timer input specified in each of its components. The 'alternative wait' is then used to deschedule the process if none of the channel or timer inputs is ready; the process will be re-scheduled when any one of them becomes ready. The channel and timer inputs are then 'disabled'. The 'disable' instructions are also designed to select the component of the alternative to be executed; the first component found to be ready is executed.

3.7   Floating point instructions

The core of the floating point instruction set was established fairly early in the design of the IMS T800. This core includes simple load, store and arithmetic instructions. Examination of statistics derived from FORTRAN programs suggested that the addition of some more complex instructions would improve performance and code density. Proposed changes to the instruction set were assesed by examining their effect on a number of numerical programs. For each proposed instruction set, a compiler was constructed, the programs compiled with it, and the resulting code then run on a simulator. The resulting instruction set is now described.

In the IMS T800 operands are transferred between the transputer's memory and the floating point evaluation stack by means of floating point load and store instructions. There are two groups of such instructions, one for single length numbers, one for double length. In the description of the load and store instructions which follow only the double length instructions are described. However, there are single length instructions which correspond with each of the double length instructions.

The address of a floating point operand is computed on the CPU's stack and the operand is then loaded, from the addressed memory location, onto the FPU's stack. Operands in the floating point stack are tagged with their length. The operand's tag will be set when the operand is loaded or is computed. The tags allow the number of instructions needed for floating point operations to be reduced; there is no need, for example, to have both *floating add single* and *floating add double* instructions; a single *floating add* will suffice.

3.7.1 Optimising use of the stack

The depth of the register stacks in the CPU and FPU is carefully chosen. Floating point expressions commonly have embedded address calculations, as the operands of floating point operators are often elements of one dimensional or two dimensional arrays. The CPU stack is deep enough to allow most integer calculations and address calculations to be performed within it. Similarly, the depth of the FPU stack allows most floating point expressions to be evaluated within it, employing the CPU stack to form addresses for the operands.

No hardware is used to deal with stack overflow. A compiler can easily examine expressions and introduce temporary variables in memory to avoid stack overflow. The number of such temporary variables can be minimised by careful choice of the evaluation order; an algorithm to perform this optimisation is given in Bibliography reference *INMOS '87*. The algorithm, already used to optimise the use of the integer stack of the IMS T414, is also used for the main CPU of the IMS T800.

3.7.2 Concurrent operation of FPU and CPU

In the IMS T800 the FPU operates concurrently with the CPU. This means that it is possible to perform an address calculation in the CPU whilst the FPU performs a floating point calculation. This can lead to significant performance improvements in real applications which access arrays heavily. This aspect of the IMS T800's performance was carefully assessed, partly through examination of the 'Livermore Loops' (Bibliography reference *McMahon*). These are a collection of small kernels designed to represent the types of calculation performed on super-computers. They are of interest because they contain constructs which occur in real programs which are not represented in such programs as the Whetstone benchmark (see below). In particular, they contain accesses to two and three-dimensional arrays, operations where the concurrency within the IMS T800 is used to good effect. In some cases the compiler is able to choose the order of performing address calculations so as to maximise overlapping; this involves a modification of the algorithm mentioned earlier.

As a simple example of overlapping consider the implementation of Livermore Loop 7. The IMS T800-30 achieves a speed of 2.25 Mflops on this benchmark; for comparison the IMS T800-20 achieves 1.5 Mflops, the IMS T414-20 achieves 0.09 Mflops and a VAX 11/780 (with floating point accelerator - fpa) achieves 0.54 Mflops. The occam program for loop 7 is as follows:

```
-- LIVERMORE LOOP 7
SEQ k = 0 FOR n
  x[k] :=    u[k] + ((( r*(z[k] + (r*y[k]))) +
             (t*((u[k+3] + (r*(u[k+2] + (r*u[k+1])))))) +
             (t*((u[k+6] + (r*(u[k+5] + (r*u[k+4])))))))
```

The first stage in the computation of this is to load the value `y[k]`. This requires a sequence of four instructions. A further three instructions cause `r` to be loaded and the FPU multiply to be initiated.

Although the floating point multiplication takes several cycles to complete, the CPU is able to continue executing instructions whilst the FPU performs the multiplication. Thus the CPU can execute the next segment of code which computes the address of `z[k]` whilst the FPU perfroms the multiplication.

Finally, the value `z[k]` is pushed onto the floating point stack and added to the previously computed subexpression `r*y[k]`. It is not until value `z[k]` is loaded that the CPU needs to synchronise with the FPU.

The computation of the remainder of the expression proceeds in the same way, and the FPU never has to wait for the CPU to perform an address calculation.

3.8 Floating point unit design

In designing a concurrent systems component such as the IMS T800, it is important to maximise the performance obtained from a given area of silicon; many components can be used together to deliver more performance. This contrasts with the design of a conventional co-processor where the aim is to maximise the performance of a single processor by the use of a large area of silicon. As a result, in designing the IMS T800, the performance benefits of silicon hungry devices such as barrel shifters and flash multipliers were carefully examined.

A flash multiplier is too large to fit on chip together with the processor, and would therefore necessitate the use of a separate co-processor chip. The introduction of a co-processor interface to a separate chip slows down the rate at which operands can be transferred to and from the floating point unit. Higher performance can, therefore, be obtained from a slow multiplier on the same chip as the processor than from a fast one on a separate chip. This leads to an important conclusion: *a separate co-processor chip is not appropriate for scalar floating point arithmetic*. A separate co-processor would be effective where a large amount of work can be handed to the co-processor by transferring a small amount of information; for example a vector co-processor would require only the addresses of its vector operands to be transferred via the co-processor interface.

It turns out that a flash multiplier also operates much more quickly than is necessary. Only a pipelined vector processor can deliver operands at a rate consistent with the use of such devices. In fact, any useful floating point calculation involves more operand accesses than operations. As an example consider the assignment `y[i] := y[i] + (t * x[i])` which constitutes the core of the LINPACK floating point benchmark. To perform this it is necessary to load three operands, perform two operations and to store a result. If we assume that it takes twice as long to perform a floating point operation as to load or store a floating point number then the execution time of this example would be evenly split between operand access time and operation time. This means that there would be at most a factor of two available in performance improvement from the use of an infinitely fast floating point unit!

Unlike a flash multiplier, a fast normalising shifter is important for fast floating point operation. When implementing IEEE arithmetic it may be necessary to perform a long shift on every floating point operation and unless a fast shifter is incorporated into the floating point unit the maximum operation time can become very long. Fortunately, unlike a flash multiplier, it is possible to design a fast shifter in a reasonable area of silicon. The shifter used in the IMS T800 is designed to perform a shift in a single cycle and to normalise in two cycles.

Consequently, the floating point unit of the IMS T800 contains a fast normalising shifter but not a flash multiplier. However there is a certain amount of logic devoted to multiplication and division. Multiplication is performed three-bits per cycle, and division is performed two-bits per cycle. This gives rise to a single length multiplication time of 13 cycles (433 nS (-30), or 650 nS (-20)) and a double length divide time of 34 cycles (1.13 μS (-30), or 1.7 μS (-20)).

Figure 3.15 illustrates the physical layout of the floating point unit.

The datapaths contain registers and shift paths. The fraction datapath is 59 bits wide, and the exponent data path is 13 bits wide. The normalising shifter interfaces to both the fraction data path and the exponent datapath. This is because the data to be shifted will come from the fraction datapath whilst the magnitude of the shift is associated with the exponent datapath. One further interesting aspect of the design is the microcode ROM. Although the diagram shows two ROMs, they are both part of the same logical ROM. This has been split in two so that control signals do not need to be bussed through the datapaths.

3.9 Floating point performance

The IMS T414 has microcode support for 32-bit floating point arithmetic which gives it performance comparable with the current generation of floating point co-processors. It achieves an operation time of about 10 microseconds on single length IEEE 754 floating point numbers. The IMS T800-20 betters the floating point operation speed of the IMS T414 by more than an order of magnitude; its operation times are shown in table 3.1.

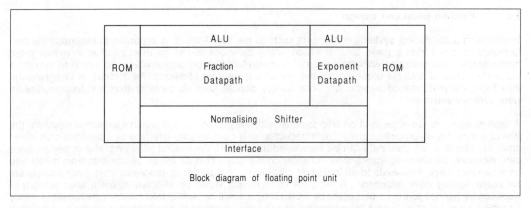

Figure 3.15: Floating point unit block diagram

Table 3.1: Floating point performance

Operation	IMS T800-30		IMS T800-20		IMS T414-20	
	Single	Double	Single	Double	Single	Double
add	233 nS	233 nS	350 nS	350 nS	11.5 μS	28.3 μS
subtract	233 nS	233 nS	350 nS	350 nS	11.5 μS	28.3 μS
multiply	367 nS	667 nS	550 nS	1000 nS	10.0 μS	38.0 μS
divide	567 nS	1067 nS	850 nS	1600 nS	12.3 μS	55.75 μS

The operation time is not a reliable measure of performance on real numerical programs. For this reason, floating point performance is often measured by the Whetstone benchmark. The Whetstone benchmark provides a good mix of floating point operations, and also includes procedure calls, array indexing and transcendental functions. It is, in some sense, a 'typical' scientific program.

The performance of the IMS T414 and IMS T800 as measured by the Whetstone benchmark is shown in table 3.2.

Another important measure is the performance obtained from a given area of silicon - or the performance of a single chip. The IMS T800 requires negligible support circuitry and can even be used without external memory. Consequently, the circuit board area needed for a typical microprocessor with co-processor, memory and support circuitry can instead be used for several IMS T800 devices, providing several times the performance in any concurrent application.

Table 3.2: Floating point performance

Processor		Whetstones/second single length
VAX 11/780 FPA	UNIX 4.3 BSD	1083K
IMS T414-20	20 MHz	663K
IMS T800-20	20 MHz	4000K
IMS T800-30	30 MHz	6000K

3 Transputer internal architecture

3.10 Graphics capability

The 'bit-blt' operations of a conventional graphics processor no longer seem appropriate in these days of byte (or greater) per pixel colour displays. The fast block move of the IMS T414 make it suitable for use in graphics applications using byte-per-pixel colour displays.

The block move on the IMS T414 is designed to saturate the memory bandwidth, moving any number of bytes from any byte boundary in memory to any other byte boundary using the smallest possible number of word read and write operations. Using the transputer's internal memory the block move sustains a transfer rate of 60 Mbytes per second (-30), or 40 Mbytes per second (-20); using the fastest possible external memory the block move sustains 20 Mbytes per second (-30) or 13.3 Mbytes per second (-20).

The IMS T800 extends this capability by incorporation of a two-dimensional version of the block move (**Move2d**) which can move windows around a screen at full memory bandwidth, and conditional versions of the same block move which can be used to place templates and text into windows. One of these operations (**Draw2d**) copies bytes from source to destination, writing only non-zero bytes to the destination. A new object of any shape can therefore be drawn on top of the current image. A further operation (**Clip2d**) copies only zero bytes in the source. All of these instructions achieve the speed of the simple IMS T414 move instruction, enabling a 1 million pixel screen to be drawn 13 times per second. Unlike the conventional 'bit-blt' instruction, it is never necessary to read the destination data.

3.10.1 Example - drawing coloured text

Drawing proportional spaced text provides a simple example of the use of the IMS T800 instructions. The font is stored in a two dimensional array **Font**; the height of **Font** is the fixed character height, and the start of each character is defined by an array **start**. The textures of the character and its background are selected from an array of textures; the textures providing a range of colours or even stripes and tartans!

An occam procedure to perform such drawing is given below and the effect of each stage in the drawing process is illustrated by the diagrams on the final page of this document. First, (1) the texture for the character is selected and copied to a temporary area and (2) the character in the font is used to clip this texture to the appropriate shape. Then (3) the background texture is selected and copied to the screen, and (4) the new character drawn on top of it.

```
-- Draw character ch in texture F on background texture B
PROC DrawChar(VAL INT Ch, F, B)
  SEQ
    IF
      (x + width[ch]) > screenwidth
        SEQ
          x := 0
          y := y + height
      (x + width[ch]) <= screenwidth
        SKIP
    [height] [maxwidth] BYTE Temp :
    SEQ
      Move2d(Texture[F],0,0, Temp,0,0, width[ch],height)
      Clip2d(Font[ch],start[ch],0, Temp,0,0, width[ch],height)
      Move2d(Texture[B],0,0, Screen,x,y, width[ch],height))
      Draw2d(Temp,0,0, Screen,x,y, width[ch],height)
      x := x + width[ch]
```

This procedure will fill a 1 million pixel screen with proportionally spaced characters in about 1/6 second. Obviously, a simpler and faster version could be used if the character colour or background colour was restricted. The operation of this procedure is illustrated in figure 3.16.

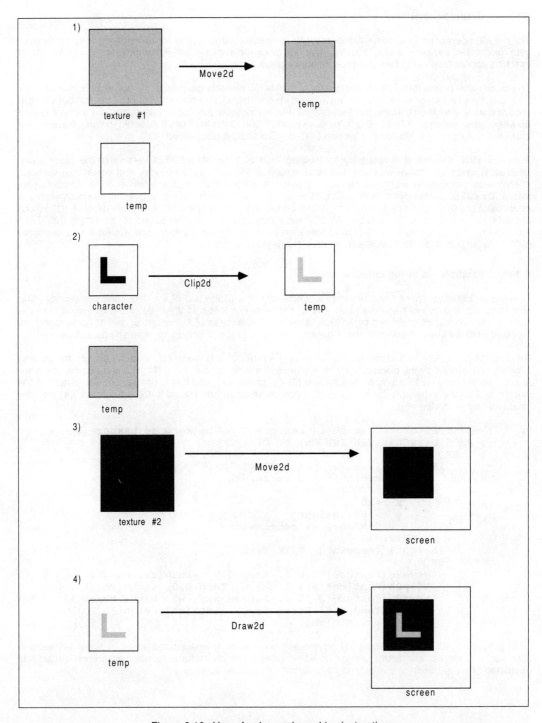

Figure 3.16: Use of enhanced graphics instructions

4   Conclusion

The INMOS transputer family is a range of system components which can be used to construct high performance concurrent systems. As all members of the family incorporate INMOS communications links, a system may be constructed from different members of the family. All transputers provide hardware support for concurrency and offer exceptional performance on process scheduling, inter-process communication and inter-transputer communication.

The design of the transputers takes careful note of silicon economics. The central processor used in the transputer offers a performance comparable with that of other VLSI processors several times larger. The small size of the processor allows a memory and communications system to be integrated on to the same VLSI device. This level of integration allows very fast access to memory and very fast inter-transputer communication. Similarly, the transputer floating point unit is integrated into the same device as the central processor, eliminating the delays inherent in communicating data between devices.

Chapter 3

IMS T800 engineering data

1 Introduction

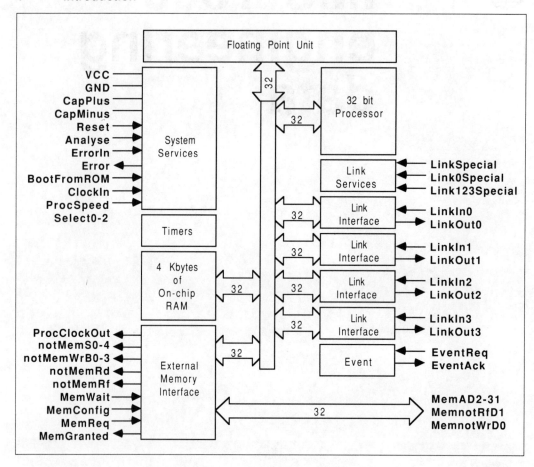

Figure 1.1: IMS T800 block diagram

1 Introduction

The IMS T800 transputer is a 32 bit CMOS microcomputer with a 64 bit floating point unit and graphics support. It has 4 Kbytes on-chip RAM for high speed processing, a configurable memory interface and four standard INMOS communication links. The instruction set achieves efficient implementation of high level languages and provides direct support for the occam model of concurrency when using either a single transputer or a network. Procedure calls, process switching and typical interrupt latency are sub-microsecond.

The processor speed of a device can be pin-selected in stages from 17.5 MHz up to the maximum allowed for the part. A device running at 30 MHz achieves an instruction throughput of 15 MIPS.

The IMS T800 provides high performance arithmetic and floating point operations. The 64 bit floating point unit provides single and double length operation to the ANSI-IEEE 754-1985 standard for floating point arithmetic. It is able to perform floating point operations concurrently with the processor, sustaining a rate of 1.5 Mflops at a processor speed of 20 MHz and 2.25 Mflops at 30 MHz.

High performance graphics support is provided by microcoded block move instructions which operate at the speed of memory. The two-dimensional block move instructions provide for contiguous block moves as well as block copying of either non-zero bytes of data only or zero bytes only. Block move instructions can be used to provide graphics operations such as text manipulation, windowing, panning, scrolling and screen updating.

Cyclic redundancy checking (CRC) instructions are available for use on arbitrary length serial data streams, to provide error detection where data integrity is critical. Another feature of the IMS T800, useful for pattern recognition, is the facility to count bits set in a word.

The IMS T800 can directly access a linear address space of 4 Gbytes. The 32 bit wide memory interface uses multiplexed data and address lines and provides a data rate of up to 4 bytes every 100 nanoseconds (40 Mbytes/sec) for a 30 MHz device. A configurable memory controller provides all timing, control and DRAM refresh signals for a wide variety of mixed memory systems.

System Services include processor reset and bootstrap control, together with facilities for error analysis. Error signals may be daisy-chained in multi-transputer systems.

The standard INMOS communication links allow networks of transputer family products to be constructed by direct point to point connections with no external logic. The IMS T800 links support the standard operating speed of 10 Mbits per second, but also operate at 5 or 20 Mbits per second. Each link can transfer data bi-directionally at up to 2.35 Mbytes/sec.

The IMS T800-20 is pin compatible with the IMS T414-20, as the extra inputs used are all held to ground on the IMS T414. The IMS T800-20 can thus be plugged directly into a circuit designed for a 20 MHz version of the IMS T414. Software should be recompiled, although no changes to the source code are necessary.

The transputer is designed to implement the occam language, detailed in the occam Reference Manual, but also efficiently supports other languages such as C, Pascal and Fortran. Access to specific features of the IMS T800is described in the relevant system development manual. Access to the transputer at machine level is seldom required, but if necessary refer to The Transputer Instruction Set - A Compiler Writers' Guide.

This data sheet supplies hardware implementation and characterisation details for the IMS T800. It is intended to be read in conjunction with the Transputer Reference Manual, which details the architecture of the transputer and gives an overview of occam.

For convenience of description, the IMS T800 operation is split into the basic blocks shown in figure 1.1.

2 Pin designations

Table 2.1: IMS T800 system services

Pin	In/Out	Function
VCC, GND		Power supply and return
CapPlus, CapMinus		External capacitor for internal clock power supply
ClockIn	in	Input clock
ProcSpeedSelect0-2	in	Processor speed selectors
Reset	in	System reset
Error	out	Error indicator
ErrorIn	in	Error daisychain input
Analyse	in	Error analysis
BootFromRom	in	Boot from external ROM or from link
HoldToGND		Must be connected to **GND**
DoNotWire		Must not be wired

Table 2.2: IMS T800 external memory interface

Pin	In/Out	Function
ProcClockOut	out	Processor clock
MemnotWrD0	in/out	Multiplexed data bit 0 and write cycle warning
MemnotRfD1	in/out	Multiplexed data bit 1 and refresh warning
MemAD2-31	in/out	Multiplexed data and address bus
notMemRd	out	Read strobe
notMemWrB0-3	out	Four byte-addressing write strobes
notMemS0-4	out	Five general purpose strobes
notMemRf	out	Dynamic memory refresh indicator
MemWait	in	Memory cycle extender
MemReq	in	Direct memory access request
MemGranted	out	Direct memory access granted
MemConfig	in	Memory configuration data input

Table 2.3: IMS T800 event

Pin	In/Out	Function
EventReq	in	Event request
EventAck	out	Event request acknowledge

Table 2.4: IMS T800 link

Pin	In/Out	Function
LinkIn0-3	in	Four serial data input channels
LinkOut0-3	out	Four serial data output channels
LinkSpecial	in	Select non-standard speed as 5 or 20 Mbits/sec
Link0Special	in	Select special speed for Link 0
Link123Special	in	Select special speed for Links 1,2,3

Signal names are prefixed by **not** if they are active low, otherwise they are active high.
Pinout details for various packages are given on page 160.

3 Processor

The 32 bit processor contains instruction processing logic, instruction and work pointers, and an operand register. It directly accesses the high speed 4 Kbyte on-chip memory, which can store data or program. Where larger amounts of memory or programs in ROM are required, the processor has access to 4 Gbytes of memory via the External Memory Interface (EMI).

3.1 Registers

The design of the transputer processor exploits the availability of fast on-chip memory by having only a small number of registers; six registers are used in the execution of a sequential process. The small number of registers, together with the simplicity of the instruction set, enables the processor to have relatively simple (and fast) data-paths and control logic. The six registers are:

> The workspace pointer which points to an area of store where local variables are kept.

> The instruction pointer which points to the next instruction to be executed.

> The operand register which is used in the formation of instruction operands.

> The A, B and C registers which form an evaluation stack.

A, B and C are sources and destinations for most arithmetic and logical operations. Loading a value into the stack pushes B into C, and A into B, before loading A. Storing a value from A, pops B into A and C into B.

Expressions are evaluated on the evaluation stack, and instructions refer to the stack implicitly. For example, the *add* instruction adds the top two values in the stack and places the result on the top of the stack. The use of a stack removes the need for instructions to respecify the location of their operands. Statistics gathered from a large number of programs show that three registers provide an effective balance between code compactness and implementation complexity.

No hardware mechanism is provided to detect that more than three values have been loaded onto the stack. It is easy for the compiler to ensure that this never happens.

Any location in memory can be accessed relative to the workpointer register, enabling the workspace to be of any size.

Further register details are given in The Transputer Instruction Set - A Compiler Writers' Guide.

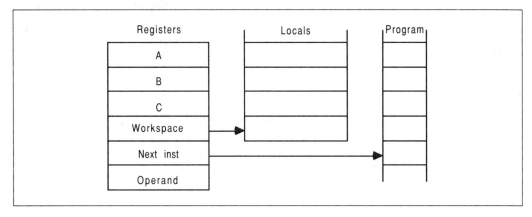

Figure 3.1: Registers

3.2 Instructions

The instruction set has been designed for simple and efficient compilation of high-level languages. All instructions have the same format, designed to give a compact representation of the operations occurring most frequently in programs.

Each instruction consists of a single byte divided into two 4-bit parts. The four most significant bits of the byte are a function code and the four least significant bits are a data value.

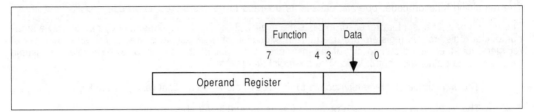

Figure 3.2: Instruction format

3.2.1 Direct functions

The representation provides for sixteen functions, each with a data value ranging from 0 to 15. Thirteen of these, shown in table 3.1, are used to encode the most important functions.

Table 3.1: Direct functions

load constant	add constant	
load local	store local	load local pointer
load non-local	store non-local	
jump	conditional jump	call

The most common operations in a program are the loading of small literal values and the loading and storing of one of a small number of variables. The *load constant* instruction enables values between 0 and 15 to be loaded with a single byte instruction. The *load local* and *store local* instructions access locations in memory relative to the workspace pointer. The first 16 locations can be accessed using a single byte instruction.

The *load non-local* and *store non-local* instructions behave similarly, except that they access locations in memory relative to the *A* register. Compact sequences of these instructions allow efficient access to data structures, and provide for simple implementations of the static links or displays used in the implementation of high level programming languages such as occam, C, Fortran, Pascal or ADA.

3.2.2 Prefix functions

Two more function codes allow the operand of any instruction to be extended in length; *prefix* and *negative prefix*.

All instructions are executed by loading the four data bits into the least significant four bits of the operand register, which is then used as the instruction's operand. All instructions except the prefix instructions end by clearing the operand register, ready for the next instruction.

The *prefix* instruction loads its four data bits into the operand register and then shifts the operand register up four places. The *negative prefix* instruction is similar, except that it complements the operand register before shifting it up. Consequently operands can be extended to any length up to the length of the operand register by a sequence of prefix instructions. In particular, operands in the range -256 to 255 can be represented using one prefix instruction.

3 Processor

The use of prefix instructions has certain beneficial consequences. Firstly, they are decoded and executed in the same way as every other instruction, which simplifies and speeds instruction decoding. Secondly, they simplify language compilation by providing a completely uniform way of allowing any instruction to take an operand of any size. Thirdly, they allow operands to be represented in a form independent of the processor wordlength.

3.2.3 Indirect functions

The remaining function code, *operate*, causes its operand to be interpreted as an operation on the values held in the evaluation stack. This allows up to 16 such operations to be encoded in a single byte instruction. However, the prefix instructions can be used to extend the operand of an *operate* instruction just like any other. The instruction representation therefore provides for an indefinite number of operations.

Encoding of the indirect functions is chosen so that the most frequently occurring operations are represented without the use of a prefix instruction. These include arithmetic, logical and comparison operations such as *add*, *exclusive or* and *greater than*. Less frequently occurring operations have encodings which require a single prefix operation.

3.2.4 Expression evaluation

Evaluation of expressions sometimes requires use of temporary variables in the workspace, but the number of these can be minimised by careful choice of the evaluation order.

Table 3.2: Expression evaluation

Program	Mnemonic
x := 0	ldc 0
	stl x
x := #24	pfix 2
	ldc 4
	stl x
x := y + z	ldl y
	ldl z
	add
	stl x

3.2.5 Efficiency of encoding

Measurements show that about 70% of executed instructions are encoded in a single byte; that is, without the use of prefix instructions. Many of these instructions, such as *load constant* and *add* require just one processor cycle.

The instruction representation gives a more compact representation of high level language programs than more conventional instruction sets. Since a program requires less store to represent it, less of the memory bandwidth is taken up with fetching instructions. Furthermore, as memory is word accessed the processor will receive four instructions for every fetch.

Short instructions also improve the effectiveness of instruction pre-fetch, which in turn improves processor performance. There is an extra word of pre-fetch buffer, so the processor rarely has to wait for an instruction fetch before proceeding. Since the buffer is short, there is little time penalty when a jump instruction causes the buffer contents to be discarded.

3.3 Processes and concurrency

A process starts, performs a number of actions, and then either stops without completing or terminates complete. Typically, a process is a sequence of instructions. A transputer can run several processes in parallel (concurrently). Processes may be assigned either high or low priority, and there may be any number of each (page 53).

The processor has a microcoded scheduler which enables any number of concurrent processes to be executed together, sharing the processor time. This removes the need for a software kernel.

At any time, a concurrent process may be

- *Active*
  - Being executed.
  - On a list waiting to be executed.

- *Inactive*
  - Ready to input.
  - Ready to output.
  - Waiting until a specified time.

The scheduler operates in such a way that inactive processes do not consume any processor time. It allocates a portion of the processor's time to each process in turn. Active processes waiting to be executed are held in two linked lists of process workspaces, one of high priority processes and one of low priority processes (page 53). Each list is implemented using two registers, one of which points to the first process in the list, the other to the last. In the Linked Process List figure 3.3, process S is executing and P, Q and R are active, awaiting execution. Only the low priority process queue registers are shown; the high priority process ones perform in a similar manner.

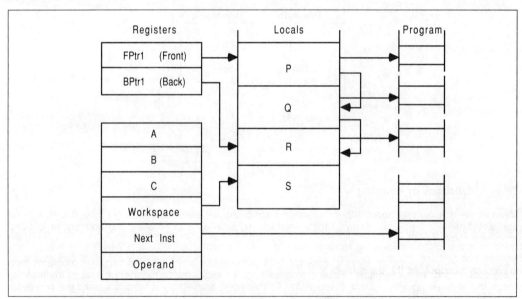

Figure 3.3: Linked process list

Table 3.3: Priority queue control registers

Function	High Priority	Low Priority
Pointer to front of active process list	*Fptr0*	*Fptr1*
Pointer to back of active process list	*Bptr0*	*Bptr1*

3 Processor 53

Each process runs until it has completed its action, but is descheduled whilst waiting for communication from another process or transputer, or for a time delay to complete. In order for several processes to operate in parallel, a low priority process is only permitted to run for a maximum of two time slices before it is forcibly descheduled at the next descheduling point (page 57). The time slice period is 5120 cycles of the external 5 MHz clock, giving ticks approximately 1ms apart.

A process can only be descheduled on certain instructions, known as descheduling points (page 57). As a result, an expression evaluation can be guaranteed to execute without the process being timesliced part way through.

Whenever a process is unable to proceed, its instruction pointer is saved in the process workspace and the next process taken from the list. Process scheduling pointers are updated by instructions which cause scheduling operations, and should not be altered directly. Actual process switch times are less than 1 μs, as little state needs to be saved and it is not necessary to save the evaluation stack on rescheduling.

The processor provides a number of special operations to support the process model, including *start process* and *end process*. When a main process executes a parallel construct, *start process* instructions are used to create the necessary additional concurrent processes. A *start process* instruction creates a new process by adding a new workspace to the end of the scheduling list, enabling the new concurrent process to be executed together with the ones already being executed. When a process is made active it is always added to the end of the list, and thus cannot pre-empt processes already on the same list.

The correct termination of a parallel construct is assured by use of the *end process* instruction. This uses a workspace location as a counter of the parallel construct components which have still to terminate. The counter is initialised to the number of components before the processes are *started*. Each component ends with an *end process* instruction which decrements and tests the counter. For all but the last component, the counter is non zero and the component is descheduled. For the last component, the counter is zero and the main process continues.

3.4 Priority

The IMS T800 supports two levels of priority. Priority 1 (low priority) processes are executed whenever there are no active priority 0 (high priority) processes.

High priority processes are expected to execute for a short time. If one or more high priority processes are able to proceed, then one is selected and runs until it has to wait for a communication, a timer input, or until it completes processing.

If no process at high priority is able to proceed, but one or more processes at low priority are able to proceed, then one is selected.

Low priority processes are periodically timesliced to provide an even distribution of processor time between computationally intensive tasks.

If there are **n** low priority processes, then the maximum latency from the time at which a low priority process becomes active to the time when it starts processing is 2**n**-2 timeslice periods. It is then able to execute for between one and two timeslice periods, less any time taken by high priority processes. This assumes that no process monopolises the transputer's time; i.e. it has a distribution of descheduling points (page 57).

Each timeslice period lasts for 5120 cycles of the external 5 MHz input clock (approximately 1 millisecond at the standard frequency of 5 MHz).

If a high priority process is waiting for an external channel to become ready, and if no other high priority process is active, then the interrupt latency (from when the channel becomes ready to when the process starts executing) is typically 19 processor cycles, a maximum of 78 cycles (assuming use of on-chip RAM). If the floating point unit is not being used at the time then the maximum interrupt latency is only 58 cycles. To ensure this latency, certain instructions are interruptable.

3.5 Communications

Communication between processes is achieved by means of channels. Process communication is point-to-point, synchronised and unbuffered. As a result, a channel needs no process queue, no message queue and no message buffer.

A channel between two processes executing on the same transputer is implemented by a single word in memory; a channel between processes executing on different transputers is implemented by point-to-point links. The processor provides a number of operations to support message passing, the most important being *input message* and *output message*.

The *input message* and *output message* instructions use the address of the channel to determine whether the channel is internal or external. Thus the same instruction sequence can be used for both, allowing a process to be written and compiled without knowledge of where its channels are connected.

The process which first becomes ready must wait until the second one is also ready. A process performs an input or output by loading the evaluation stack with a pointer to a message, the address of a channel, and a count of the number of bytes to be transferred, and then executing an *input message* or *output message* instruction. Data is transferred if the other process is ready. If the channel is not ready or is an external one the process will deschedule.

3.6 Timers

The transputer has two 32 bit timer clocks which 'tick' periodically. The timers provide accurate process timing, allowing processes to deschedule themselves until a specific time.

One timer is accessible only to high priority processes and is incremented every microsecond, cycling completely in approximately 4295 milliseconds. The other is accessible only to low priority processes and is incremented every 64 microseconds, giving exactly 15625 ticks in one second. It has a full period of approximately 76 hours.

Table 3.4: Timer registers

Clock0	Current value of high priority (level 0) process clock
Clock1	Current value of low priority (level 1) process clock
TNextReg0	Indicates time of earliest event on high priority (level 0) timer queue
TNextReg1	Indicates time of earliest event on low priority (level 1) timer queue

The current value of the processor clock can be read by executing a *load timer* instruction. A process can arrange to perform a *timer input*, in which case it will become ready to execute after a specified time has been reached. The *timer input* instruction requires a time to be specified. If this time is in the 'past' then the instruction has no effect. If the time is in the 'future' then the process is descheduled. When the specified time is reached the process is scheduled again.

Figure 3.4 shows two processes waiting on the timer queue, one waiting for time 21, the other for time 31.

3 Processor

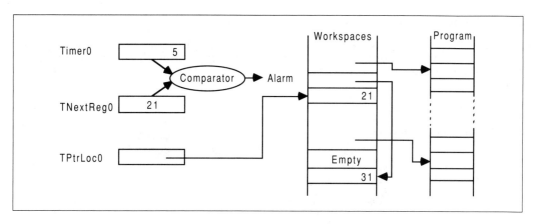

Figure 3.4: Timer registers

4 Instruction set summary

The Function Codes table 4.8. gives the basic function code set (page 50). Where the operand is less than 16, a single byte encodes the complete instruction. If the operand is greater than 15, one prefix instruction (*pfix*) is required for each additional four bits of the operand. If the operand is negative the first prefix instruction will be *nfix*.

Table 4.1: *prefix* coding

Mnemonic		Function code	Memory code
ldc	#3	#4	#43
ldc	#35		
is coded as			
pfix	#3	#2	#23
ldc	#5	#4	#45
ldc	#987		
is coded as			
pfix	#9	#2	#29
pfix	#8	#2	#28
ldc	#7	#4	#47
ldc	-31 (*ldc*	#FFFFFFE1)	
is coded as			
nfix	#1	#6	#61
ldc	#1	#4	#41

Tables 4.9 to 4.13 give details of the operation codes. Where an operation code is less than 16 (e.g. *add*: operation code **05**), the operation can be stored as a single byte comprising the *operate* function code **F** and the operand (**5** in the example). Where an operation code is greater than 15 (e.g. *ladd*: operation code **16**), the *prefix* function code **2** is used to extend the instruction.

Table 4.2: *operate* coding

Mnemonic		Function code	Memory code
add	(op. code #5)		#F5
is coded as			
opr	add	#F	#F5
ladd	(op. code #16)		#21F6
is coded as			
pfix	#1	#2	#21
opr	#6	#F	#F6

In the Floating Point Operation Codes tables 4.21 to 4.27, a selector sequence code (page 65) is indicated in the Memory Code column by **s**. The code given in the Operation Code column is the indirection code, the operand for the *ldc* instruction.

The FPU and processor operate concurrently, so the actual throughput of floating point instructions is better than that implied by simply adding up the instruction times. For full details see The Transputer Instruction Set - A Compiler Writers' Guide.

4 Instruction set summary

The Processor Cycles column refers to the number of periods **TPCLPCL** taken by an instruction executing in internal memory. The number of cycles is given for the basic operation only; where relevant the time for the *prefix* function (one cycle) should be added. For a 20 MHz transputer one cycle is 50ns. Some instruction times vary. Where a letter is included in the cycles column it is interpreted from table 4.3.

Table 4.3: Instruction set interpretation

Ident	Interpretation
b	Bit number of the highest bit set in register *A*. Bit 0 is the least significant bit.
m	Bit number of the highest bit set in the absolute value of register *A*. Bit 0 is the least significant bit.
n	Number of places shifted.
w	Number of words in the message. Part words are counted as full words. If the message is not word aligned the number of words is increased to include the part words at either end of the message.
p	Number of words per row.
r	Number of rows.

The **DE** column of the tables indicates the descheduling/error features of an instruction as described in table 4.4.

Table 4.4: Instruction features

Ident	Feature	See page:
D	The instruction is a descheduling point	57
E	The instruction will affect the *Error* flag	58, 72
F	The instruction will affect the *FP_Error* flag	65, 58

4.1 Descheduling points

The instructions in table 4.5 are the only ones at which a process may be descheduled (page 52). They are also the ones at which the processor will halt if the **Analyse** pin is asserted (page 71).

Table 4.5: Descheduling point instructions

input message	output message	output byte	output word
timer alt wait	timer input	stop on error	alt wait
jump	loop end	end process	stop process

4.2 Error instructions

The instructions in table 4.6 are the only ones which can affect the *Error* flag (page 72) directly. Note, however, that the floating point unit error flag *FP_Error* is set by certain floating point instructions (page 58), and that *Error* can be set from this flag by *fpcheckerror*.

Table 4.6: Error setting instructions

add	add constant	subtract	
multiply	fractional multiply	divide	remainder
long add	long subtract	long divide	
set error	testerr	fpcheckerror	
check word	check subscript from 0	check single	check count from 1

4.3 Floating point errors

The instructions in table 4.7 are the only ones which can affect the floating point error flag *FP_Error* (page 65). *Error* is set from this flag by *fpcheckerror* if *FP_Error* is set.

Table 4.7: Floating point error setting instructions

fpadd	fpsub	fpmul	fpdiv
fpldnladdsn	fpldnladddb	fpldnlmulsn	fpldnlmuldb
fpremfirst	fpusqrtfirst	fpgt	fpeq
fpuseterror	fpuclearerror	fptesterror	
fpuexpincby32	fpuexpdecby32	fpumulby2	fpudivby2
fpur32tor64	fpur64tor32	fpucki32	fpucki64
fprtoi32	fpuabs	fpint	

4 Instruction set summary

Table 4.8: IMS T800 function codes

Function Code	Memory Code	Mnemonic	Processor Cycles	Name	D E
0	0X	j	3	jump	D
1	1X	ldlp	1	load local pointer	
2	2X	pfix	1	prefix	
3	3X	ldnl	2	load non-local	
4	4X	ldc	1	load constant	
5	5X	ldnlp	1	load non-local pointer	
6	6X	nfix	1	negative prefix	
7	7X	ldl	2	load local	
8	8X	adc	1	add constant	E
9	9X	call	7	call	
A	AX	cj	2	conditional jump (not taken)	
			4	conditional jump (taken)	
B	BX	ajw	1	adjust workspace	
C	CX	eqc	2	equals constant	
D	DX	stl	1	store local	
E	EX	stnl	2	store non-local	
F	FX	opr	-	operate	

Table 4.9: IMS T800 arithmetic/logical operation codes

Operation Code	Memory Code	Mnemonic	Processor Cycles	Name	D E
46	24F6	and	1	and	
4B	24FB	or	1	or	
33	23F3	xor	1	exclusive or	
32	23F2	not	1	bitwise not	
41	24F1	shl	$n+2$	shift left	
40	24F0	shr	$n+2$	shift right	
05	F5	add	1	add	E
0C	FC	sub	1	subtract	E
53	25F3	mul	38	fractional multiply (no rounding)	E
72	27F2	fmul	35	fractional multiply (rounding)	E
			40	multiply	E
2C	22FC	div	39	divide	E
1F	21FF	rem	37	remainder	E
09	F9	gt	2	greater than	
04	F4	diff	1	difference	
52	25F2	sum	1	sum	
08	F8	prod	$b+4$	product for positive register A	
			$m+5$	product for negative register A	

Table 4.10: IMS T800 long arithmetic operation codes

Operation Code	Memory Code	Mnemonic	Processor Cycles	Name	DE
16	21F6	ladd	2	long add	E
38	23F8	lsub	2	long subtract	E
37	23F7	lsum	2	long sum	
4F	24FF	ldiff	2	long diff	
31	23F1	lmul	33	long multiply	
1A	21FA	ldiv	35	long divide	E
36	23F6	lshl	$n+3$	long shift left ($n<32$)	
			$n-28$	long shift left ($n\geq32$)	
35	23F5	lshr	$n+3$	long shift right ($n<32$)	
			$n-28$	long shift right ($n\geq32$)	
19	21F9	norm	$n+5$	normalise ($n<32$)	
			$n-26$	normalise ($n\geq32$)	
			3	normalise ($n=64$)	

Table 4.11: IMS T800 general operation codes

Operation Code	Memory Code	Mnemonic	Processor Cycles	Name	DE
00	F0	rev	1	reverse	
3A	23FA	xword	4	extend to word	
56	25F6	cword	5	check word	E
1D	21FD	xdble	2	extend to double	
4C	24FC	csngl	3	check single	E
42	24F2	mint	1	minimum integer	

Table 4.12: IMS T800 block move operation codes

Operation Code	Memory Code	Mnemonic	Processor Cycles	Name	DE
5B	25FB	move2dinit	8	initialise data for 2D block move	
5C	25FC	move2dall	$(2p+23)*r$	2D block copy	
5D	25FD	move2dnonzero	$(2p+23)*r$	2D block copy non-zero bytes	
5E	25FE	move2dzero	$(2p+23)*r$	2D block copy zero bytes	

Table 4.13: IMS T800 CRC and bit operation codes

Operation Code	Memory Code	Mnemonic	Processor Cycles	Name	DE
74	27F4	crcword	35	calculate crc on word	
75	27F5	crcbyte	11	calculate crc on byte	
76	27F6	bitcnt	$b+2$	count bits set in word	
77	27F7	bitrevword	36	reverse bits in word	
78	27F8	bitrevnbits	$n+4$	reverse bottom n bits in byte	

4 Instruction set summary

Table 4.14: IMS T800 indexing/array operation codes

Operation Code	Memory Code	Mnemonic	Processor Cycles	Name	DE
02	F2	bsub	1	byte subscript	
0A	FA	wsub	2	word subscript	
81	28F1	wsubdb	3	form double word subscript	
34	23F4	bcnt	2	byte count	
3F	23FF	wcnt	5	word count	
01	F1	lb	5	load byte	
3B	23FB	sb	4	store byte	
4A	24FA	move	2w+8	move message	

Table 4.15: IMS T800 timer handling operation codes

Operation Code	Memory Code	Mnemonic	Processor Cycles	Name	DE
22	22F2	ldtimer	2	load timer	
2B	22FB	tin	30	timer input (time future)	D
			4	timer input (time past)	D
4E	24FE	talt	4	timer alt start	
51	25F1	taltwt	15	timer alt wait (time past)	D
			48	timer alt wait (time future)	D
47	24F7	enbt	8	enable timer	
2E	22FE	dist	23	disable timer	

Table 4.16: IMS T800 input/output operation codes

Operation Code	Memory Code	Mnemonic	Processor Cycles	Name	DE
07	F7	in	2w+19	input message	D
0B	FB	out	2w+19	output message	D
0F	FF	outword	23	output word	D
0E	FE	outbyte	23	output byte	D
12	21F2	resetch	3	reset channel	
43	24F3	alt	2	alt start	
44	24F4	altwt	5	alt wait (channel ready)	D
			17	alt wait (channel not ready)	D
45	24F5	altend	4	alt end	
49	24F9	enbs	3	enable skip	
30	23F0	diss	4	disable skip	
48	24F8	enbc	7	enable channel (ready)	
			5	enable channel (not ready)	
2F	22FF	disc	8	disable channel	

Table 4.17: IMS T800 control operation codes

Operation Code	Memory Code	Mnemonic	Processor Cycles	Name	DE
20	22F0	ret	5	return	
1B	21FB	ldpi	2	load pointer to instruction	
3C	23FC	gajw	2	general adjust workspace	
5A	25FA	dup	1	duplicate top of stack	
06	F6	gcall	4	general call	
21	22F1	lend	10	loop end (loop)	D
			5	loop end (exit)	D

Table 4.18: IMS T800 scheduling operation codes

Operation Code	Memory Code	Mnemonic	Processor Cycles	Name	DE
0D	FD	startp	12	start process	D
03	F3	endp	13	end process	D
39	23F9	runp	10	run process	
15	21F5	stopp	11	stop process	
1E	21FE	ldpri	1	load current priority	

Table 4.19: IMS T800 error handling operation codes

Operation Code	Memory Code	Mnemonic	Processor Cycles	Name	DE
13	21F3	csub0	2	check subscript from 0	E
4D	24FD	ccnt1	3	check count from 1	E
29	22F9	testerr	2	test error false and clear (no error)	
			3	test error false and clear (error)	
10	21F0	seterr	1	set error	E
55	25F5	stoperr	2	stop on error (no error)	D
57	25F7	clrhalterr	1	clear halt-on-error	
58	25F8	sethalterr	1	set halt-on-error	
59	25F9	testhalterr	2	test halt-on-error	

Table 4.20: IMS T800 processor initialisation operation codes

Operation Code	Memory Code	Mnemonic	Processor Cycles	Name	DE
2A	22FA	testpranal	2	test processor analysing	
3E	23FE	saveh	4	save high priority queue registers	
3D	23FD	savel	4	save low priority queue registers	
18	21F8	sthf	1	store high priority front pointer	
50	25F0	sthb	1	store high priority back pointer	
1C	21FC	stlf	1	store low priority front pointer	
17	21F7	stlb	1	store low priority back pointer	
54	25F4	sttimer	1	store timer	

4 Instruction set summary

Table 4.21: IMS T800 floating point load/store operation codes

Operation Code	Memory Code	Mnemonic	Processor Cycles	Name	DE
8E	28FE	fpldnlsn	2	fp load non-local single	
8A	28FA	fpldnldb	3	fp load non-local double	
86	28F6	fpldnlsni	4	fp load non-local indexed single	
82	28F2	fpldnldbi	6	fp load non-local indexed double	
9F	29FF	fpldzerosn	2	load zero single	
A0	2AF0	fpldzerodb	2	load zero double	
AA	2AFA	fpldnladdsn	8/11	fp load non local & add single	F
A6	2AF6	fpldnladddb	9/12	fp load non local & add double	F
AC	2AFC	fpldnlmulsn	13/20	fp load non local & multiply single	F
A8	2AF8	fpldnlmuldb	21/30	fp load non local & multiply double	F
88	28F8	fpstnlsn	2	fp store non-local single	
84	28F4	fpstnldb	3	fp store non-local double	
9E	29FE	fpstnli32	4	store non-local int32	

Processor cycles are shown as **Typical/Maximum** cycles.

Table 4.22: IMS T800 floating point general operation codes

Operation Code	Memory Code	Mnemonic	Processor Cycles	Name	DE
AB	2AFB	fpentry	1	floating point unit entry	
A4	2AF4	fprev	1	fp reverse	
A3	2AF3	fpdup	1	fp duplicate	

Table 4.23: IMS T800 floating point rounding operation codes

Operation Code	Memory Code	Mnemonic	Processor Cycles	Name	DE
22	s	fpurn	1	set rounding mode to round nearest	
06	s	fpurz	1	set rounding mode to round zero	
04	s	fpurp	1	set rounding mode to round positive	
05	s	fpurm	1	set rounding mode to round minus	

Table 4.24: IMS T800 floating point error operation codes

Operation Code	Memory Code	Mnemonic	Processor Cycles	Name	DE
83	28F3	fpchkerror	1	check fp error	E
9C	29FC	fptesterror	2	test fp error false and clear	F
23	s	fpuseterror	1	set fp error	F
9C	s	fpuclearerror	1	clear fp error	F

Table 4.25: IMS T800 floating point comparison operation codes

Operation Code	Memory Code	Mnemonic	Processor Cycles	Name	DE
94	29F4	fpgt	4/6	fp greater than	F
95	29F5	fpeq	3/5	fp equality	F
92	29F2	fpordered	3/4	fp orderability	
91	29F1	fpnan	2/3	fp NaN	
93	29F3	fpnotfinite	2/2	fp not finite	
0E	s	fpuchki32	3/4	check in range of type int32	F
0F	s	fpuchki64	3/4	check in range of type int64	F

Processor cycles are shown as **Typical/Maximum** cycles.

Table 4.26: IMS T800 floating point conversion operation codes

Operation Code	Memory Code	Mnemonic	Processor Cycles	Name	DE
07	s	fpur32tor64	3/4	real32 to real64	F
08	s	fpur64tor32	6/9	real64 to real32	F
9D	29FD	fprtoi32	7/9	real to int32	F
96	29F6	fpi32tor32	8/10	int32 to real32	
98	29F8	fpi32tor64	8/10	int32 to real64	
9A	29FA	fpb32tor64	8/8	bit32 to real64	
0D	s	fpunoround	2/2	real64 to real32, no round	
A1	2AF1	fpint	5/6	round to floating integer	F

Processor cycles are shown as **Typical/Maximum** cycles.

Table 4.27: IMS T800 floating point arithmetic operation codes

Operation Code	Memory Code	Mnemonic	Processor cycles		Name	DE
			Single	Double		
87	28F7	fpadd	6/9	6/9	fp add	F
89	28F9	fpsub	6/9	6/9	fp subtract	F
8B	28FB	fpmul	11/18	18/27	fp multiply	F
8C	28FC	fpdiv	16/28	31/43	fp divide	F
0B	s	fpuabs	2/2	2/2	fp absolute	F
8F	28FF	fpremfirst	36/46	36/46	fp remainder first step	F
90	29F0	fpremstep	32/36	32/36	fp remainder iteration	
01	s	fpusqrtfirst	27/29	27/29	fp square root first step	F
02	s	fpusqrtstep	42/42	42/42	fp square root step	
03	s	fpusqrtlast	8/9	8/9	fp square root end	
0A	s	fpuexpinc32	6/9	6/9	multiply by 2^{32}	F
09	s	fpuexpdec32	6/9	6/9	divide by 2^{32}	F
12	s	fpumulby2	6/9	6/9	multiply by 2.0	F
11	s	fpudivby2	6/9	6/9	divide by 2.0	F

Processor cycles are shown as **Typical/Maximum** cycles.

5 Floating point unit

The 64 bit FPU provides single and double length arithmetic to floating point standard ANSI-IEEE 754-1985. It is able to perform floating point arithmetic concurrently with the central processor unit (CPU), sustaining in excess of 2.25 Mflops on a 30 MHz device. All data communication between memory and the FPU occurs under control of the CPU.

The FPU consists of a microcoded computing engine with a three deep floating point evaluation stack for manipulation of floating point numbers. These stack registers are *FA*, *FB* and *FC*, each of which can hold either 32 bit or 64 bit data; an associated flag, set when a floating point value is loaded, indicates which. The stack behaves in a similar manner to the CPU stack (page 49).

As with the CPU stack, the FPU stack is not saved when rescheduling (page 52) occurs. The FPU can be used in both low and high priority processes. When a high priority process interrupts a low priority one the FPU state is saved inside the FPU. The CPU will service the interrupt immediately on completing its current operation. The high priority process will not start, however, before the FPU has completed its current operation.

Points in an instruction stream where data need to be transferred to or from the FPU are called *synchronisation points*. At a synchronisation point the first processing unit to become ready will wait until the other is ready. The data transfer will then occur and both processors will proceed concurrently again. In order to make full use of concurrency, floating point data source and destination addresses can be calculated by the CPU whilst the FPU is performing operations on a previous set of data. Device performance is thus optimised by minimising the CPU and FPU idle times.

The FPU has been designed to operate on both single length (32 bit) and double length (64 bit) floating point numbers, and returns results which fully conform to the ANSI-IEEE 754-1985 floating point arithmetic standard. Denormalised numbers are fully supported in the hardware. All rounding modes defined by the standard are implemented, with the default being round to nearest.

The basic addition, subtraction, multiplication and division operations are performed by single instructions. However, certain less frequently used floating point instructions are selected by a value in register *A* (when allocating registers, this should be taken into account). A *load constant* instruction *ldc* is used to load register *A*; the *floating point entry* instruction *fpentry* then uses this value to select the floating point operation. This pair of instructions is termed a *selector sequence*.

Names of operations which use *fpentry* begin with *fpu*. A typical usage, returning the absolute value of a floating point number, would be

> ldc   fpuabs;   fpentry;

Since the indirection code for *fpuabs* is **0B**, it would be encoded as

Table 5.1: *fpentry* coding

Mnemonic		Function code	Memory code
ldc	fpuabs	#4	#4B
fpentry **is coded as**	(op. code #AB)		#2AFB
pfix	#A	#2	#2A
opr	#B	#F	#FB

The *remainder* and *square root* instructions take considerably longer than other instructions to complete. In order to minimise the interrupt latency period of the transputer they are split up to form instruction sequences. As an example, the instruction sequence for a single length square root is

fpusqrtfirst;   fpusqrtstep;   fpusqrtstep;   fpusqrtlast;

The FPU has its own error flag *FP_Error*. This reflects the state of evaluation within the FPU and is set in circumstances where invalid operations, division by zero or overflow exceptions to the ANSI-IEEE 754-1985standard would be flagged (page 58). *FP_Error* is also set if an input to a floating point operation is infinite or is not a number (NaN). The *FP_Error* flag can be set, tested and cleared without affecting the main *Error* flag, but can also set *Error* when required (page 58). Depending on how a program is compiled, it is possible for both unchecked and fully checked floating point arithmetic to be performed.

Further details on the operation of the FPU can be found in The Transputer Instruction Set - A Compiler Writers' Guide.

Table 5.2: Typical floating point operation times for IMS T800

Operation	T800-20		T800-30	
	Single length	Double length	Single length	Double length
add	350 ns	350 ns	233 ns	233 ns
subtract	350 ns	350 ns	233 ns	233 ns
multiply	550 ns	1000 ns	367 ns	667 ns
divide	850 ns	1600 ns	567 ns	1067 ns

Timing is for operations where both operands are normalised fp numbers.

6 System services

System services include all the necessary logic to initialise and sustain operation of the device. They also include error handling and analysis facilities.

6.1 Power

Power is supplied to the device via the **VCC** and **GND** pins. Several of each are provided to minimise inductance within the package. All supply pins must be connected. The supply must be decoupled close to the chip by at least one 100nF low inductance (e.g. ceramic) capacitor between **VCC** and **GND**. Four layer boards are recommended; if two layer boards are used, extra care should be taken in decoupling.

Input voltages must not exceed specification with respect to **VCC** and **GND**, even during power-up and power-down ramping, otherwise *latchup* can occur. CMOS devices can be permanently damaged by excessive periods of latchup.

6.2 CapPlus, CapMinus

The internally derived power supply for internal clocks requires an external low leakage, low inductance $1\,\mu$F capacitor to be connected between **CapPlus** and **CapMinus**. A ceramic capacitor is preferred, with an impedance less than 3 ohms between 100 KHz and 10 MHz. If a polarised capacitor is used the negative terminal should be connected to **CapMinus**. Total PCB track length should be less than 50mm. The connections must not touch power supplies or other noise sources.

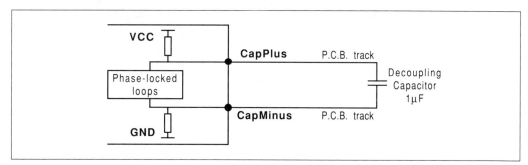

Figure 6.1: Recommended PLL decoupling

6.3 ClockIn

Transputer family components use a standard clock frequency, supplied by the user on the **ClockIn** input. The nominal frequency of this clock for all transputer family components is 5MHz, regardless of device type, transputer word length or processor cycle time. High frequency internal clocks are derived from **ClockIn**, simplifying system design and avoiding problems of distributing high speed clocks externally.

A number of transputer devices may be connected to a common clock, or may have individual clocks providing each one meets the specified stability criteria. In a multi-clock system the relative phasing of **ClockIn** clocks is not important, due to the asynchronous nature of the links. Mark/space ratio is unimportant provided the specified limits of **ClockIn** pulse widths are met.

Oscillator stability is important. **ClockIn** must be derived from a crystal oscillator; RC oscillators are not sufficiently stable. **ClockIn** must not be distributed through a long chain of buffers. Clock edges must be monotonic and remain within the specified voltage and time limits.

Table 6.1: Input clock

SYMBOL	PARAMETER	MIN	NOM	MAX	UNITS	NOTE
TDCLDCH	ClockIn pulse width low	40			ns	
TDCHDCL	ClockIn pulse width high	40			ns	
TDCLDCL	ClockIn period		200		ns	1,3
TDCerror	ClockIn timing error			±0.5	ns	2
TDC1DC2	Difference in ClockIn for 2 linked devices			400	ppm	3
TDCr	ClockIn rise time			10	ns	4
TDCf	ClockIn fall time			8	ns	4

Notes

1 Measured between corresponding points on consecutive falling edges.

2 Variation of individual falling edges from their nominal times.

3 This value allows the use of 200ppm crystal oscillators for two devices connected together by a link.

4 Clock transitions must be monotonic within the range **VIH** to **VIL** (page 99).

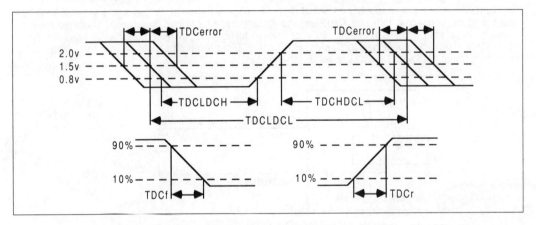

Figure 6.2: ClockIn timing

6.4   ProcSpeedSelect0-2

Processor speed of the IMS T800 is variable in discrete steps. The desired speed can be selected, up to the maximum rated for a particular component, by the three speed select lines **ProcSpeedSelect0-2**. The pins are tied high or low, according to the table below, for the various speeds. The **ProcSpeedSelect0-2** pins are designated **HoldToGND** on the IMS T414, and coding is so arranged that the IMS T800 can be plugged directly into a board designed for a 20MHz IMS T414.

Only six of the possible speed select combinations are currently used; the other two are not valid speed selectors. The frequency of **ClockIn** for the speeds given in the table is 5 MHz.

6 System services

Table 6.2: Processor speed selection

Proc Speed Select2	Proc Speed Select1	Proc Speed Select0	Processor Clock Speed MHz	Processor Cycle Time nS	Notes
0	0	0	20.0	50.0	
0	0	1	22.5	44.4	
0	1	0	25.0	40.0	
0	1	1	30.0	33.3	
1	0	0	35.0	28.6	
1	0	1			Invalid
1	1	0	17.5	57.1	
1	1	1			Invalid

Note: Inclusion of a speed selection in this table does not imply immediate availability.

6.5 Reset

Reset can go high with **VCC**, but must at no time exceed the maximum specified voltage for **VIH**. After **VCC** is valid **ClockIn** should be running for a minimum period **TDCVRL** before the end of **Reset**. The falling edge of **Reset** initialises the transputer, triggers the memory configuration sequence and starts the bootstrap routine. Link outputs are forced low during reset; link inputs and **EventReq** should be held low. Memory request (DMA) must not occur whilst **Reset** is high but can occur before bootstrap (page 93).

After the end of **Reset** there will be a delay of 144 periods of **ClockIn** (figure 6.3). Following this, the **MemWrD0**, **MemRfD1** and **MemAD2-31** pins will be scanned to check for the existence of a pre-programmed memory interface configuration (page 84). This lasts for a further 144 periods of **ClockIn**. Regardless of whether a configuration was found, 36 configuration read cycles will then be performed on external memory using the default memory configuration (page 85), in an attempt to access the external configuration ROM. A delay will then occur, its period depending on the actual configuration. Finally eight complete and consecutive refresh cycles will initialise any dynamic RAM, using the new memory configuration. If the memory configuration does not enable refresh of dynamic RAM the refresh cycles will be replaced by an equivalent delay with no external memory activity.

If **BootFromRom** is high bootstrapping will then take place immediately, using data from external memory; otherwise the transputer will await an input from any link. The processor will be in the low priority state.

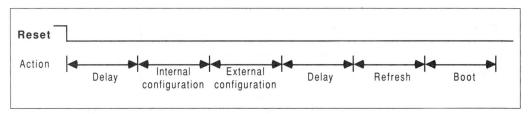

Figure 6.3: IMS T800 post-reset sequence

6.6 Bootstrap

The transputer can be bootstrapped either from a link or from external ROM. To facilitate debugging, **BootFromRom** may be dynamically changed but must obey the specified timing restrictions.

If **BootFromRom** is connected high (e.g. to **VCC**) the transputer starts to execute code from the top two bytes in external memory, at address #7FFFFFFE. This location should contain a backward jump to a program in ROM. The processor is in the low priority state, and the *W* register points to *MemStart* (page 73).

Table 6.3: Reset and Analyse

SYMBOL	PARAMETER	MIN	NOM	MAX	UNITS	NOTE
TPVRH	Power valid before Reset	10			ms	
TRHRL	Reset pulse width high	8			ClockIn	1
TDCVRL	ClockIn running before Reset end	10			ms	2
TAHRH	Analyse setup before Reset	3			ms	
TRLAL	Analyse hold after Reset end	1			ns	
TBRVRL	BootFromRom setup	0			ms	
TRLBRX	BootFromRom hold after Reset	50			ms	
TALBRX	BootFromRom hold after Analyse	50			ms	

Notes

1 Full periods of **ClockIn TDCLDCL** required.

2 At power-on reset.

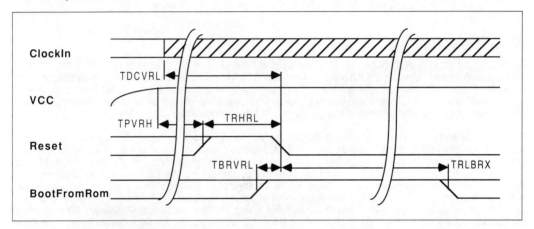

Figure 6.4: Transputer reset timing with Analyse low

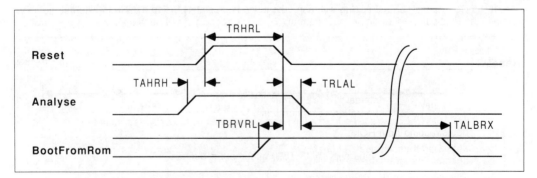

Figure 6.5: Transputer reset and analyse timing

6 System services

If **BootFromRom** is connected low (e.g. to **GND**) the transputer will wait for the first bootstrap message to arrive on any one of its links. The transputer is ready to receive the first byte on a link within two processor cycles **TPCLPCL** after **Reset** goes low.

If the first byte received (the control byte) is greater than 1 it is taken as the quantity of bytes to be input. The following bytes, to that quantity, are then placed in internal memory starting at location *MemStart*. Following reception of the last byte the transputer will start executing code at *MemStart* as a low priority process. The memory space immediately above the loaded code is used as work space. Messages arriving on other links after the control byte has been received and on the bootstrapping link after the last bootstrap byte will be retained until a process inputs from them.

6.7 Peek and poke

Any location in internal or external memory can be interrogated and altered when the transputer is waiting for a bootstrap from link. If the control byte is 0 then eight more bytes are expected on the same link. The first four byte word is taken as an internal or external memory address at which to poke (write) the second four byte word. If the control byte is 1 the next four bytes are used as the address from which to peek (read) a word of data; the word is sent down the output channel of the same link.

Following such a peek or poke, the transputer returns to its previously held state. Any number of accesses may be made in this way until the control byte is greater than 1, when the transputer will commence reading its bootstrap program. Any link can be used, but addresses and data must be transmitted via the same link as the control byte.

6.8 Analyse

If **Analyse** is taken high when the transputer is running, the transputer will halt at the next descheduling point (page 57). From **Analyse** being asserted, the processor will halt within three time slice periods plus the time taken for any high priority process to complete. As much of the transputer status is maintained as is necessary to permit analysis of the halted machine. Memory refresh continues.

Input links will continue with outstanding transfers. Output links will not make another access to memory for data but will transmit only those bytes already in the link buffer. Providing there is no delay in link acknowledgement, the links should be inactive within a few microseconds of the transputer halting.

Reset should not be asserted before the transputer has halted and link transfers have ceased. When **Reset** is taken low whilst **Analyse** is high, neither the memory configuration sequence nor the block of eight refresh cycles will occur; the previous memory configuration will be used for any external memory accesses. If **BootFromRom** is high the transputer will bootstrap as soon as **Analyse** is taken low, otherwise it will await a control byte on any link. If **Analyse** is taken low without **Reset** going high the transputer state and operation are undefined. After the end of a valid **Analyse** sequence the registers have the values given in table 6.4.

Table 6.4: Register values after analyse

I	*MemStart* if bootstrapping from a link, or the external memory bootstrap address if bootstrapping from ROM.
W	*MemStart* if bootstrapping from ROM, or the address of the first free word after the bootstrap program if bootstrapping from link.
A	The value of *I* when the processor halted.
B	The value of *W* when the processor halted, together with the priority of the process when the transputer was halted (i.e. the *W* descriptor).
C	The ID of the bootstrapping link if bootstrapping from link.

6.9 Error, ErrorIn

The **Error** pin carries the OR'ed output of the internal *Error* flag and the **ErrorIn** input. If **Error** is high it indicates either that **ErrorIn** is high or that an error was detected in one of the processes. An internal error can be caused, for example, by arithmetic overflow, divide by zero, array bounds violation or software setting the flag directly (page 58). It can also be set from the floating point unit under certain circumstances (page 58, 65). Once set, the *Error* flag is only cleared by executing the instruction *testerr* (page 56). The error is not cleared by processor reset, in order that analysis can identify any errant transputer (page 71).

A process can be programmed to stop if the *Error* flag is set; it cannot then transmit erroneous data to other processes, but processes which do not require that data can still be scheduled. Eventually all processes which rely, directly or indirectly, on data from the process in error will stop through lack of data. **ErrorIn** does not directly affect the status of a processor in any way.

By setting the *HaltOnError* flag the transputer itself can be programmed to halt if *Error* becomes set. If *Error* becomes set after *HaltOnError* has been set, all processes on that transputer will cease but will not necessarily cause other transputers in a network to halt. Setting *HaltOnError* after *Error* will not cause the transputer to halt; this allows the processor reset and analyse facilities to function with the flags in indeterminate states.

An alternative method of error handling is to have the errant process or transputer cause all transputers to halt. This can be done by 'daisy-chaining' the **ErrorIn** and **Error** pins of a number of processors and applying the final **Error** output signal to the **EventReq** pin of a suitably programmed master transputer. Since the process state is preserved when stopped by an error, the master transputer can then use the analyse function to debug the fault. When using such a circuit, note that the *Error* flag is in an indeterminate state on power up; the circuit and software should be designed with this in mind.

Error checks can be removed completely to optimise the performance of a proven program; any unexpected error then occurring will have an arbitrary undefined effect.

If a high priority process pre-empts a low priority one, status of the *Error* and *HaltOnError* flags is saved for the duration of the high priority process and restored at the conclusion of it. Status of both flags is transmitted to the high priority process. Either flag can be altered in the process without upsetting the error status of any complex operation being carried out by the pre-empted low priority process.

In the event of a transputer halting because of *HaltOnError*, the links will finish outstanding transfers before shutting down. If **Analyse** is asserted then all inputs continue but outputs will not make another access to memory for data. Memory refresh will continue to take place.

After halting due to the *Error* flag changing from 0 to 1 whilst *HaltOnError* is set, register *I* points two bytes past the instruction which set *Error*. After halting due to the **Analyse** pin being taken high, register *I* points one byte past the instruction being executed. In both cases *I* will be copied to register *A*.

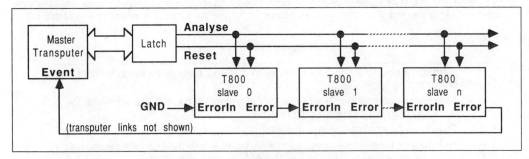

Figure 6.6: Error handling in a multi-transputer system

7   Memory

The IMS T800 has 4 Kbytes of fast internal static memory for high rates of data throughput. Each internal memory access takes one processor cycle **ProcClockOut** (page 75). The transputer can also access 4 Gbytes of external memory space. Internal and external memory are part of the same linear address space.

IMS T800 memory is byte addressed, with words aligned on four-byte boundaries. The least significant byte of a word is the lowest addressed byte.

The bits in a byte are numbered 0 to 7, with bit 0 the least significant. The bytes are numbered from 0, with byte 0 the least significant. In general, wherever a value is treated as a number of component values, the components are numbered in order of increasing numerical significance, with the least significant component numbered 0. Where values are stored in memory, the least significant component value is stored at the lowest (most negative) address.

Internal memory starts at the most negative address #80000000 and extends to #80000FFF. User memory begins at #80000070; this location is given the name *MemStart*.

A reserved area at the bottom of internal memory is used to implement link and event channels.

Two words of memory are reserved for timer use, *TPtrLoc0* for high priority processes and *TPtrLoc1* for low priority processes. They either indicate the relevant priority timer is not in use or point to the first process on the timer queue at that priority level.

Values of certain processor registers for the current low priority process are saved in the reserved *IntSaveLoc* locations when a high priority process pre-empts a low priority one. Other locations are reserved for extended features such as block moves and floating point operations.

External memory space starts at #80001000 and extends up through #00000000 to #7FFFFFFF. Memory configuration data and ROM bootstrapping code must be in the most positive address space, starting at #7FFFFF6C and #7FFFFFFE respectively. Address space immediately below this is conventionally used for ROM based code.

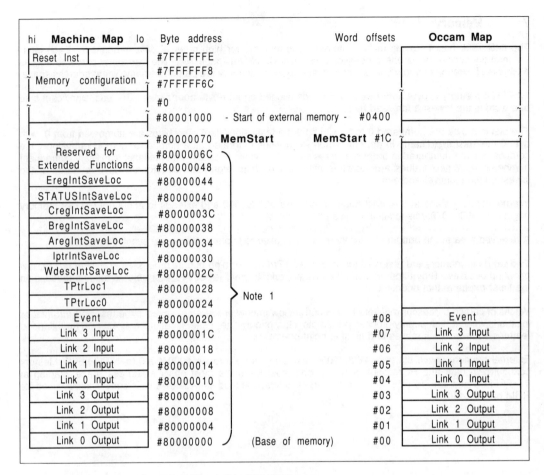

Figure 7.1: IMS T800 memory map

These locations are used as auxiliary processor registers and should not be manipulated by the user. Like processor registers, their contents may be useful for implementing debugging tools (**Analyse**, page 71). For details see The Transputer Instruction Set - A Compiler Writers' Guide.

8 External memory interface

The External Memory Interface (EMI) allows access to a 32 bit address space, supporting dynamic and static RAM as well as ROM and EPROM. EMI timing can be configured at **Reset** to cater for most memory types and speeds, and a program is supplied with the Transputer Development System to aid in this configuration.

There are 13 internal configurations which can be selected by a single pin connection (page 84). If none are suitable the user can configure the interface to specific requirements, as shown in page 85.

8.1 ProcClockOut

This clock is derived from the internal processor clock, which is in turn derived from **ClockIn**. Its period is equal to one internal microcode cycle time, and can be derived from the formula

$$TPCLPCL = TDCLDCL / PLLx$$

where **TPCLPCL** is the **ProcClockOut Period**, **TDCLDCL** is the **ClockIn Period** and **PLLx** is the phase lock loop factor for the relevant speed part, obtained from the ordering details (Ordering appendix).

The time value **Tm** is used to define the duration of **Tstates** and, hence, the length of external memory cycles; its value is exactly half the period of one **ProcClockOut** cycle (0.5∗**TPCLPCL**), regardless of mark/space ratio of **ProcClockOut**.

Edges of the various external memory strobes coincide with rising or falling edges of **ProcClockOut**. It should be noted, however, that there is a skew associated with each coincidence. The value of skew depends on whether coincidence occurs when the **ProcClockOut** edge and strobe edge are both rising, when both are falling or if either is rising when the other is falling. Timing values given in the strobe tables show the best and worst cases. If a more accurate timing relationship is required, the exact **Tstate** timing and strobe edge to **ProcClockOut** relationships should be calculated and the correct skew factors applied from the edge skew timing table 8.4.

8.2 Tstates

The external memory cycle is divided into six **Tstates** with the following functions:

- T1 Address setup time before address valid strobe.
- T2 Address hold time after address valid strobe.
- T3 Read cycle tristate or write cycle data setup.
- T4 Extendable data setup time.
- T5 Read or write data.
- T6 Data hold.

Under normal conditions each **Tstate** may be from one to four periods **Tm** long, the duration being set during memory configuration. The default condition on **Reset** is that all **Tstates** are the maximum four periods **Tm** long to allow external initialisation cycles to read slow ROM.

Period **T4** can be extended indefinitely by adding externally generated wait states.

An external memory cycle is always an even number of periods **Tm** in length and the start of **T1** always coincides with a rising edge of **ProcClockOut**. If the total configured quantity of periods **Tm** is an odd number, one extra period **Tm** will be added at the end of **T6** to force the start of the next **T1** to coincide with a rising edge of **ProcClockOut**. This period is designated **E** in configuration diagrams (page 85).

Table 8.1: ProcClockOut

SYMBOL	PARAMETER	MIN	NOM	MAX	UNITS	NOTE
TPCLPCL	ProcClockOut period	a-1	a	a+1	ns	1
TPCHPCL	ProcClockOut pulse width high	b-2.5	b	b+2.5	ns	2
TPCLPCH	ProcClockOut pulse width low		c		ns	3
Tm	ProcClockOut half cycle	b-0.5	b	b+0.5	ns	2
TPCstab	ProcClockOut stability			4	%	4

Notes

1 **a** is **TDCLDCL/PLLx**.

2 **b** is 0.5∗**TPCLPCL** (half the processor clock period).

3 **c** is **TPCLPCL-TPCHPCL**.

4 Stability is the variation of cycle periods between two consecutive cycles, measured at corresponding points on the cycles.

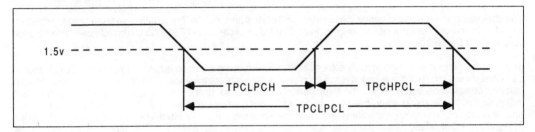

Figure 8.1: IMS T800 ProcClockOut timing

8.3   Internal access

During an internal memory access cycle the external memory interface bus **MemAD2-31** reflects the word address used to access internal RAM, **MemnotWrD0** reflects the read/write operation and **MemnotRfD1** is high; all control strobes are inactive. This is true unless and until a memory refresh cycle or DMA (memory request) activity takes place, when the bus will carry the appropriate external address or data.

The bus activity is not adequate to trace the internal operation of the transputer in full, but may be used for hardware debugging in conjuction with peek and poke (page 71).

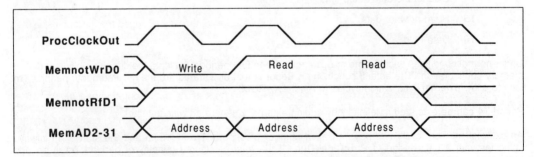

Figure 8.2: IMS T800 bus activity for internal memory cycle

8   External memory interface

8.4   MemAD2-31

External memory addresses and data are multiplexed on one bus. Only the top 30 bits of address are output on the external memory interface, using pins **MemAD2-31**. They are normally output only during **Tstates T1** and **T2**, and should be latched during this time. Byte addressing is carried out internally by the transputer for read cycles. For write cycles the relevant bytes in memory are addressed by the write strobes **notMemWrB0-3**.

The data bus is 32 bits wide. It uses **MemAD2-31** for the top 30 bits and **MemnotRfD1** and **MemnotWrD0** for the lower two bits. Read cycle data may be set up on the bus at any time after the start of **T3**, but must be valid when the transputer reads it at the end of **T5**. Data may be removed any time during **T6**, but must be off the bus no later than the end of that period.

Write data is placed on the bus at the start of **T3** and removed at the end of **T6**. If **T6** is extended to force the next cycle **Tmx** (page 77) to start on a rising edge of **ProcClockOut**, data will be valid during this time also.

8.5   MemnotWrD0

During **T1** and **T2** this pin will be low if the cycle is a write cycle, otherwise it will be high. During **Tstates T3** to **T6** it becomes bit 0 of the data bus. In both cases it follows the general timing of **MemAD2-31**.

8.6   MemnotRfD1

During **T1** and **T2**, this pin is low if the address on **MemAD2-31** is a refresh address, otherwise it is high. During **Tstates T3** to **T6** it becomes bit 1 of the data bus. In both cases it follows the general timing of **MemAD2-31**.

8.7   notMemRd

For a read cycle the read strobe **notMemRd** is low during **T4** and **T5**. Data is read by the transputer on the rising edge of this strobe, and may be removed immediately afterward. If the strobe duration is insufficient it may be extended by adding extra periods **Tm** to either or both of the **Tstates T4** and **T5**. Further extension may be obtained by inserting wait states at the end of **T4**.

In the read cycle timing diagrams **ProcClockOut** is included as a guide only; it is shown with each **Tstate** configured to one period **Tm**.

8.8   notMemS0-4

To facilitate control of different types of memory and devices, the EMI is provided with five strobe outputs, four of which can be configured by the user. The strobes are conventionally assigned the functions shown in the read and write cycle diagrams, although there is no compulsion to retain these designations.

notMemS0 is a fixed format strobe. Its leading edge is always coincident with the start of **T2** and its trailing edge always coincident with the end of **T5**.

The leading edge of **notMemS1** is always coincident with the start of **T2**, but its duration may be configured to be from zero to 31 periods **Tm**. Regardless of the configured duration, the strobe will terminate no later than the end of **T6**. The strobe is sometimes programmed to extend beyond the normal end of **Tmx**. When wait states are inserted into an EMI cycle the end of **Tmx** is delayed, but the potential active duration of the strobe is not altered. Thus the strobe can be configured to terminate relatively early under certain conditions (page 91). If **notMemS1** is configured to be zero it will never go low.

notMemS2, **notMemS3** and **notMemS4** are identical in operation. They all terminate at the end of **T5**, but the start of each can be delayed from one to 31 periods **Tm** beyond the start of **T2**. If the duration of one of these strobes would take it past the end of **T5** it will stay high. This can be used to cause a strobe to become active only when wait states are inserted. If one of these strobes is configured to zero it will never go high. Figure 8.5 shows the effect of **Wait** on strobes in more detail; each division on the scale is one period **Tm**.

Table 8.2: Read

SYMBOL	PARAMETER	MIN	NOM	MAX	UNITS	NOTE
TaZdV	Address tristate to data valid	0			ns	
TdVRdH	Data setup before read	20			ns	
TRdHdX	Data hold after read	0			ns	
TSOLRdL	notMemS0 before start of read	a-2	a	a+2	ns	1
TSOHRdH	End of read from end of notMemS0	-1		1	ns	
TRdLRdH	Read period	b		b+6	ns	2

Notes

1 **a** is total of **T2**+**T3** where **T2**, **T3** can be from one to four periods **Tm** each in length.

2 **b** is total of **T4**+**Twait**+**T5** where **T4**, **T5** can be from one to four periods **Tm** each in length and **Twait** may be any number of periods **Tm** in length.

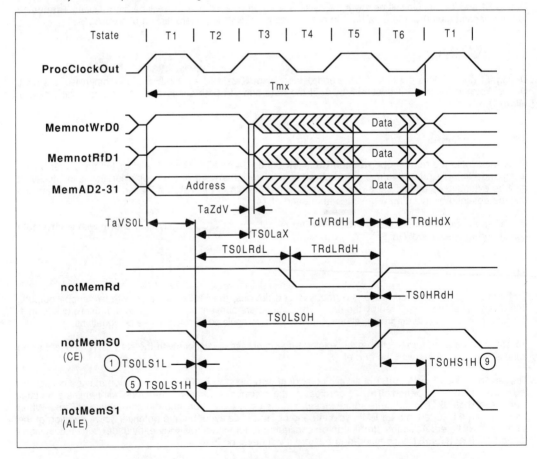

Figure 8.3: IMS T800 external read cycle: static memory

8 External memory interface

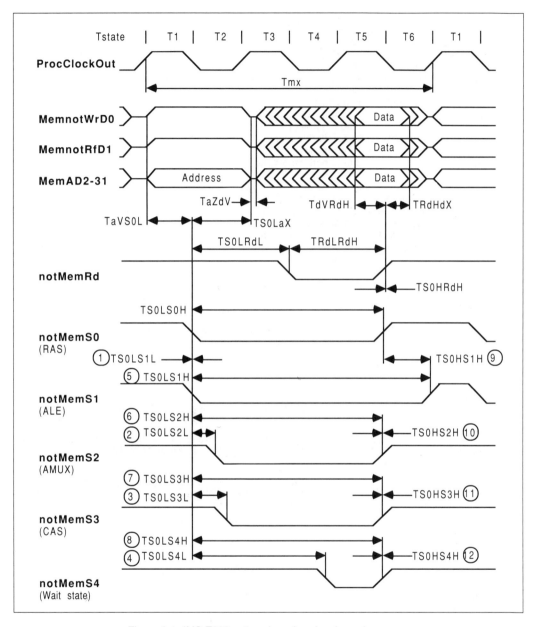

Figure 8.4: IMS T800 external read cycle: dynamic memory

3 IMS T800 engineering data

Table 8.3: IMS T800 strobe timing

SYMBOL	(n)	PARAMETER	MIN	NOM	MAX	UNITS	NOTE
TaVS0L		Address setup before notMemS0		a		ns	1
TS0LaX		Address hold after notMemS0		b		ns	2
TS0LS0H		notMemS0 pulse width low	c		c+6	ns	3
TS0LS1L	1	notMemS1 from notMemS0	0		2	ns	
TS0LS1H	5	notMemS1 end from notMemS0	d		d+6	ns	4,6
TS0HS1H	9	notMemS1 end from notMemS0 end	e-1		e+4	ns	5,6
TS0LS2L	2	notMemS2 delayed after notMemS0	f-1		f+4	ns	7
TS0LS2H	6	notMemS2 end from notMemS0	c+4		c+8	ns	3
TS0HS2H	10	notMemS2 end from notMemS0 end	0		2	ns	
TS0LS3L	3	notMemS3 delayed after notMemS0	f-1		f+3	ns	7
TS0LS3H	7	notMemS3 end from notMemS0	c+4		c+8	ns	3
TS0HS3H	11	notMemS3 end from notMemS0 end	0		2	ns	
TS0LS4L	4	notMemS4 delayed after notMemS0	f-1		f+2	ns	7
TS0LS4H	8	notMemS4 end from notMemS0	c+4		c+8	ns	3
TS0HS4H	12	notMemS4 end from notMemS0 end	0		2	ns	
Tmx		Complete external memory cycle		g			8

Notes

1 **a** is **T1** where **T1** can be from one to four periods **Tm** in length.

2 **b** is **T2** where **T2** can be from one to four periods **Tm** in length.

3 **c** is total of **T2**+**T3**+**T4**+**Twait**+**T5** where **T2**, **T3**, **T4**, **T5** can be from one to four periods **Tm** each in length and **Twait** may be any number of periods **Tm** in length.

4 **d** can be from zero to 31 periods **Tm** in length.

5 **e** can be from -27 to +4 periods **Tm** in length.

6 If the configuration would cause the strobe to remain active past the end of **T6** it will go high at the end of **T6**. If the strobe is configured to zero periods **Tm** it will remain high throughout the complete cycle **Tmx**.

7 **f** can be from zero to 31 periods **Tm** in length. If this length would cause the strobe to remain active past the end of **T5** it will go high at the end of **T5**. If the strobe value is zero periods **Tm** it will remain low throughout the complete cycle **Tmx**.

8 **g** is one complete external memory cycle comprising the total of **T1**+**T2**+**T3**+**T4**+**Twait**+**T5**+**T6** where **T1**, **T2**, **T3**, **T4**, **T5** can be from one to four periods **Tm** each in length, **T6** can be from one to five periods **Tm** in length and **Twait** may be zero or any number of periods **Tm** in length.

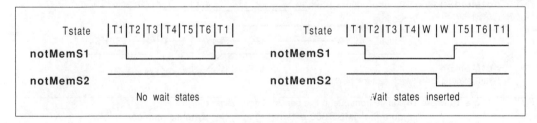

Figure 8.5: IMS T800 effect of wait states on strobes

8 External memory interface

Table 8.4: Strobe S0 to ProcClockOut skew

SYMBOL	PARAMETER	MIN	NOM	MAX	UNITS	NOTE
TPCHS0H	Strobe rising from ProcClockOut rising	0		3	ns	
TPCLS0H	Strobe rising from ProcClockOut falling	1		4	ns	
TPCHS0L	Strobe falling from ProcClockOut rising	-3		0	ns	
TPCLS0L	Strobe falling from ProcClockOut falling	-1		2	ns	

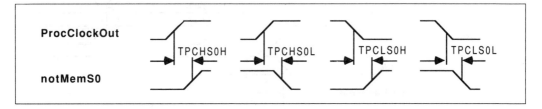

Figure 8.6: IMS T800 skew of notMemS0 to ProcClockOut

8.9 notMemWrB0-3

Because the transputer uses word addressing, four write strobes are provided; one to write each byte of the word. **notMemWrB0** addresses the least significant byte.

The transputer has both early and late write cycle modes. For a late write cycle the relevant write strobes **notMemWrB0-3** are low during **T4** and **T5**; for an early write they are also low during **T3**. Data should be latched into memory on the rising edge of the strobes in both cases, although it is valid until the end of **T6**. If the strobe duration is insufficient, it may be extended at configuration time by adding extra periods **Tm** to either or both of **Tstates T4** and **T5** for both early and late modes. For an early cycle they may also be added to **T3**. Further extension may be obtained by inserting wait states at the end of **T4**. If the data hold time is insufficient, extra periods **Tm** may be added to **T6** to extend it.

Table 8.5: Write

SYMBOL	PARAMETER	MIN	NOM	MAX	UNITS	NOTE
TdVWrH	Data setup before write	d			ns	1,5
TWrHdX	Data hold after write	a			ns	1,2
TS0LWrL	notMemS0 before start of early write	b-3		b+2	ns	1,3
	notMemS0 before start of late write	c-3		c+2	ns	1,4
TS0HWrH	End of write from end of notMemS0	-2		2	ns	1
TWrLWrH	Early write pulse width	d		d+6	ns	1,5
	Late write pulse width	e		e+6	ns	1,6

Notes

1 Timing is for all write strobes **notMemWrB0-3**.

2 **a** is **T6** where **T6** can be from one to five periods **Tm** in length.

3 **b** is **T2** where **T2** can be from one to four periods **Tm** in length.

4 **c** is total of **T2+T3** where **T2**, **T3** can be from one to four periods **Tm** each in length.

5 **d** is total of **T3+T4+Twait+T5** where **T3**, **T4**, **T5** can be from one to four periods **Tm** each in length and **Twait** may be zero or any number of periods **Tm** in length.

6 **e** is total of **T4+Twait+T5** where **T4**, **T5** can be from one to four periods **Tm** each in length and **Twait** may be zero or any number of periods **Tm** in length.

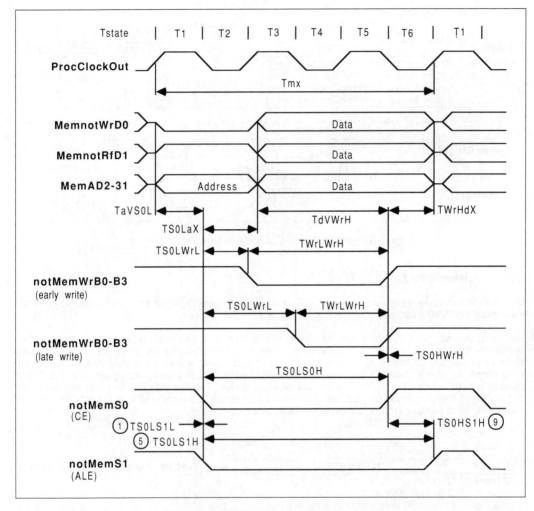

Figure 8.7: IMS T800 external write cycle

In the write cycle timing diagram **ProcClockOut** is included as a guide only; it is shown with each **Tstate** configured to one period **Tm**. The strobe is inactive during internal memory cycles.

8 External memory interface

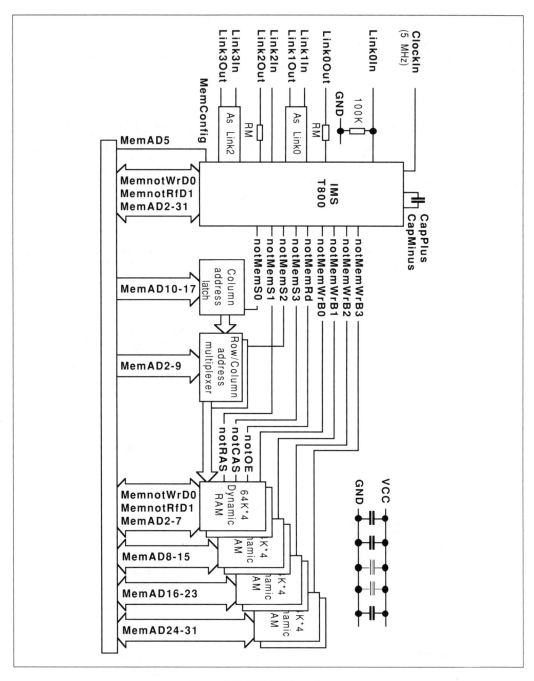

Figure 8.8: IMS T800 application

8.10 MemConfig

MemConfig is an input pin used to read configuration data when setting external memory interface (EMI) characteristics. It is read by the processor on two occasions after **Reset** goes low; first to check if one of the preset internal configurations is required, then to determine a possible external configuration.

8.10.1 Internal configuration

The internal configuration scan comprises 64 periods **TDCLDCL** of **ClockIn** during the internal scan period of 144 **ClockIn** periods. **MemnotWrD0**, **MemnotRfD1** and **MemAD2-32** are all high at the beginning of the scan. Starting with **MemnotWrD0**, each of these lines goes low successively at intervals of two **ClockIn** periods and stays low until the end of the scan. If one of these lines is connected to **MemConfig** the preset internal configuration mode associated with that line will be used as the EMI configuration. The default configuration is that defined in the table for **MemAD31**; connecting **MemConfig** to **VCC** will also produce this default configuration. Note that only 13 of the possible configurations are valid.

Table 8.6: IMS T800 internal configuration coding

Pin	Duration of each Tstate periods Tm						Strobe coefficient				Write cycle type	Refresh interval ClockIn cycles	Cycle time Proc cycles	Extra cycles e
	T1	T2	T3	T4	T5	T6	s1	s2	s3	s4				
MemnotWrD0	1	1	1	1	1	1	30	1	3	5	late	72	3	2
MemnotRfD1	1	2	1	1	1	2	30	1	2	7	late	72	4	3
MemAD2	1	2	1	1	2	3	30	1	2	7	late	72	5	4
MemAD3	2	3	1	1	2	3	30	1	3	8	late	72	6	5
MemAD4	1	1	1	1	1	1	3	1	2	3	early	72	3	2
MemAD5	1	1	2	1	2	1	5	1	2	3	early	72	4	3
MemAD6	2	1	2	1	3	1	6	1	2	3	early	72	5	4
MemAD7	2	2	2	1	3	2	7	1	3	4	early	72	6	5
MemAD8	1	1	1	1	1	1	30	1	2	3	early	—	3	2
MemAD9	1	1	2	1	2	1	30	2	5	9	early	—	4	3
MemAD10	2	2	2	2	4	2	30	2	3	8	late	72	7	6
MemAD11	3	3	3	3	3	3	30	2	4	13	late	72	9	8
MemAD31	4	4	4	4	4	4	31	30	30	18	late	72	12	11

Table 8.7: IMS T800 internal configuration description

Pin	Configuration
MemnotWrD0	Dynamic RAM in 3 processor cycles
MemnotRfD1	Dynamic RAM in 4 processor cycles
MemAD2	Dynamic RAM in 5 processor cycles
MemAD3	Dynamic RAM in 6 cycles
MemAD4	Multiplexed address dynamic RAM in 3 processor cycles
MemAD5	Multiplexed address dynamic RAM in 4 processor cycles
MemAD6	Multiplexed address dynamic RAM in 5 processor cycles
MemAD7	Multiplexed address dynamic RAM in 6 processor cycles
MemAD8	Fast static RAM in 3 processor cycles
MemAD9	Static RAM in 4 cycles with wait generator
MemAD10	General purpose configuration in 7 processor cycles
MemAD11	General purpose configuration in 9 processor cycles
MemAD31	General purpose configuration in 12 processor cycles

8 External memory interface

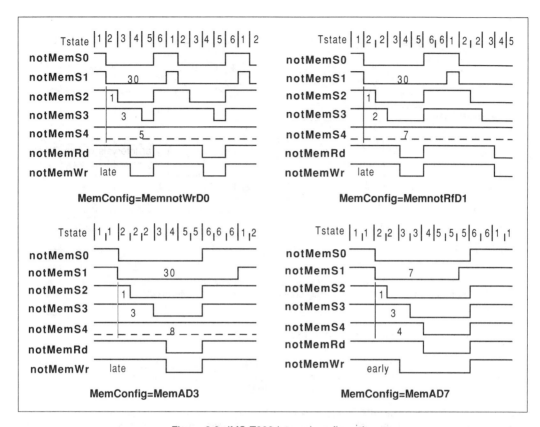

Figure 8.9: IMS T800 internal configuration

8.10.2 External configuration

If **MemConfig** is held low until **MemnotWrD0** goes low the internal configuration is ignored and an external configuration will be loaded instead. An external configuration scan always follows an internal one, but if an internal configuration occurs any external configuration is ignored.

The external configuration scan comprises 36 successive external read cycles, using the default EMI configuration preset by **MemAD31**. However, instead of data being read on the data bus as for a normal read cycle, only a single bit of data is read on **MemConfig** at each cycle. Addresses put out on the bus for each read cycle are shown in table 8.8, and are designed to address ROM at the top of the memory map. The table shows the data to be held in ROM; data required at the **MemConfig** pin is the inverse of this.

MemConfig is typically connected via an inverter to **MemnotWrD0**. Data bit zero of the least significant byte of each ROM word then provides the configuration data stream. By switching **MemConfig** between various data bus lines up to 32 configurations can be stored in ROM, one per bit of the data bus. **MemConfig** can be permanently connected to a data line or to **GND**. Connecting **MemConfig** to **GND** gives all **Tstates** configured to four periods; **notMemS1** pulse of maximum duration; **notMemS2-4** delayed by maximum; refresh interval 72 periods of **ClockIn**; refresh enabled; late write.

The external memory configuration table 8.8 shows the contribution of each memory address to the 13 configuration fields. The lowest 12 words (#7FFFFF6C to #7FFFFF98, fields 1 to 6) define the number of extra periods **Tm** to be added to each **Tstate**. If field 2 is 3 then three extra periods will be added to **T2** to extend it to the maximum of four periods.

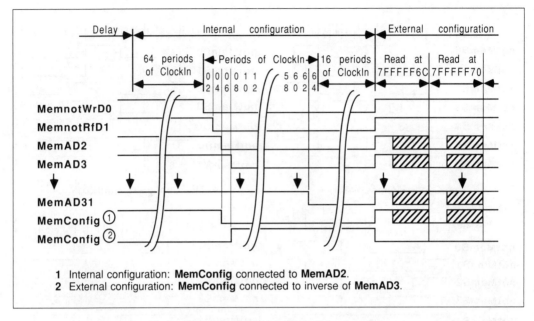

Figure 8.10: IMS T800 internal configuration scan

The next five addresses (field 7) define the duration of **notMemS1** and the following fifteen (fields 8 to 10) define the delays before strobes **notMemS2-4** become active. The five bits allocated to each strobe allow durations of from 0 to 31 periods **Tm**, as described in strobes page 77.

Addresses #7FFFFFEC to #7FFFFFF4 (fields 11 and 12) define the refresh interval and whether refresh is to be used, whilst the final address (field 13) supplies a high bit to **MemConfig** if a late write cycle is required.

The columns to the right of the coding table show the values of each configuration bit for the four sample external configuration diagrams. Note the inclusion of period **E** at the end of **T6** in some diagrams. This is inserted to bring the start of the next **Tstate T1** to coincide with a rising edge of **ProcClockOut** (page 75).

Wait states **W** have been added to show the effect of them on strobe timing; they are not part of a configuration. In each case which includes wait states, two wait periods are defined. This shows that if a wait state would cause the start of **T5** to coincide with a falling edge of **ProcClockOut**, another period **Tm** is generated by the EMI to force it to coincide with a rising edge of **ProcClockOut**. This coincidence is only necessary if wait states are added, otherwise coincidence with a falling edge is permitted.

8 External memory interface

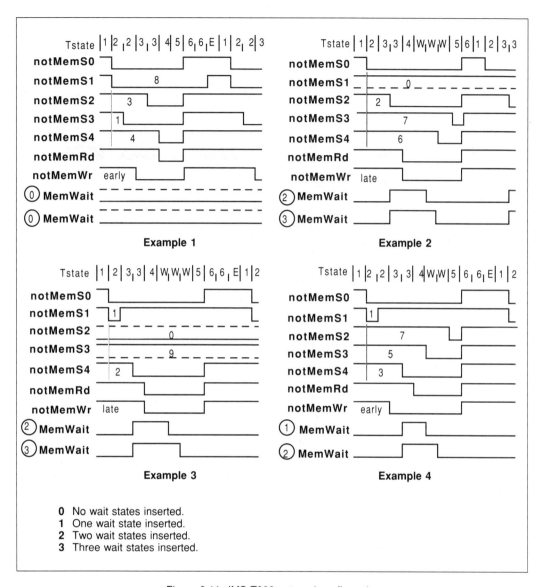

Figure 8.11: IMS T800 external configuration

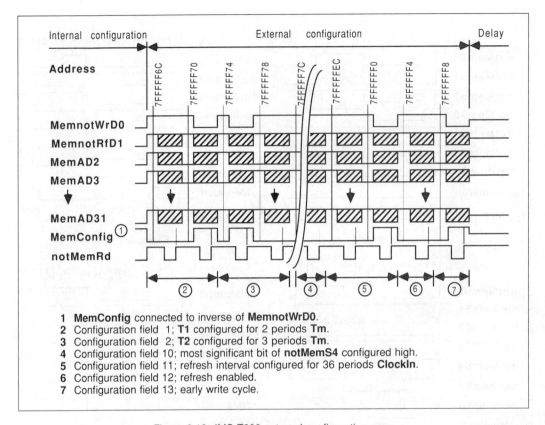

Figure 8.12: IMS T800 external configuration scan

8 External memory interface

Table 8.8: IMS T800 external configuration coding

Scan cycle	MemAD address	Field	Function	Example diagram 1	2	3	4
1	7FFFFF6C	1	T1 least significant bit	0	0	0	0
2	7FFFFF70	1	T1 most significant bit	0	0	0	0
3	7FFFFF74	2	T2 least significant bit	1	0	0	1
4	7FFFFF78	2	T2 most significant bit	0	0	0	0
5	7FFFFF7C	3	T3 least significant bit	1	1	1	1
6	7FFFFF80	3	T3 most significant bit	0	0	0	0
7	7FFFFF84	4	T4 least significant bit	0	0	0	0
8	7FFFFF88	4	T4 most significant bit	0	0	0	0
9	7FFFFF8C	5	T5 least significant bit	0	0	0	0
10	7FFFFF90	5	T5 most significant bit	0	0	0	0
11	7FFFFF94	6	T6 least significant bit	1	0	1	1
12	7FFFFF98	6	T6 most significant bit	0	0	0	0
13	7FFFFF9C	7	notMemS1 least significant bit	0	0	1	1
14	7FFFFFA0	7		0	0	0	0
15	7FFFFFA4	7	⇓ ⇓	0	0	0	0
16	7FFFFFA8	7		1	0	0	0
17	7FFFFFAC	7	notMemS1 most significant bit	0	0	0	0
18	7FFFFFB0	8	notMemS2 least significant bit	1	0	0	1
19	7FFFFFB4	8		1	1	0	1
20	7FFFFFB8	8	⇓ ⇓	0	0	0	1
21	7FFFFFBC	8		0	0	0	0
22	7FFFFFC0	8	notMemS2 most significant bit	0	0	0	0
23	7FFFFFC4	9	notMemS3 least significant bit	1	1	1	1
24	7FFFFFC8	9		0	1	0	0
25	7FFFFFCC	9	⇓ ⇓	0	1	0	1
26	7FFFFFD0	9		0	0	1	0
27	7FFFFFD4	9	notMemS3 most significant bit	0	0	0	0
28	7FFFFFD8	10	notMemS4 least significant bit	0	0	0	1
29	7FFFFFDC	10		0	1	1	1
30	7FFFFFE0	10	⇓ ⇓	1	1	0	0
31	7FFFFFE4	10		0	0	0	0
32	7FFFFFE8	10	notMemS4 most significant bit	0	0	0	0
33	7FFFFFEC	11	Refresh Interval least significant bit	-	-	-	-
34	7FFFFFF0	11	Refresh Interval most significant bit	-	-	-	-
35	7FFFFFF4	12	Refresh Enable	-	-	-	-
36	7FFFFFF8	13	Late Write	0	1	1	0

Table 8.9: IMS T800 memory refresh configuration coding

Refresh interval	Interval in μs	Field 11 encoding	Complete cycle (mS)
18	3.6	00	0.922
36	7.2	01	1.843
54	10.8	10	2.765
72	14.4	11	3.686

Refresh intervals are in periods of **ClockIn** and **ClockIn** frequency is 5MHz:

$$\text{Interval} = 18 * 200 = 3600\text{ns}$$

Refresh interval is between successive incremental refresh addresses.
Complete cycles are shown for 256 row DRAMS.

Table 8.10: Memory configuration

SYMBOL	PARAMETER	MIN	NOM	MAX	UNITS	NOTE
TMCVRdH	Memory configuration data setup	20			ns	
TRdHMCX	Memory configuration data hold	0			ns	
TS0LRdH	notMemS0 to configuration data read	a		a+6	ns	1

Notes

1 **a** is 16 periods **Tm**.

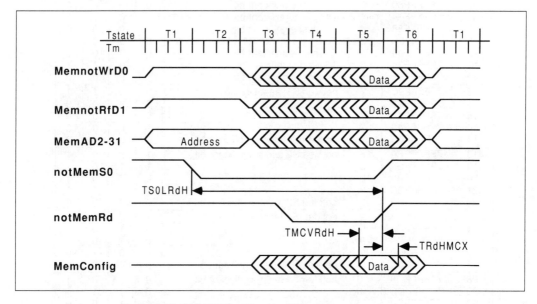

Figure 8.13: IMS T800 external configuration read cycle timing

8.11   notMemRf

The IMS T800 can be operated with memory refresh enabled or disabled. The selection is made during memory configuration, when the refresh interval is also determined. Refresh cycles do not interrupt internal memory accesses, although the internal addresses cannot be reflected on the external bus during refresh.

When refresh is disabled no refresh cycles occur. During the post-**Reset** period eight dummy refresh cycles will occur with the appropriate timing but with no bus or strobe activity.

A refresh cycle uses the same basic external memory timing as a normal external memory cycle, except that it starts two periods **Tm** before the start of **T1**. If a refresh cycle is due during an external memory access, it will be delayed until the end of that external cycle. Two extra periods **Tm** (periods **R** in the diagram) will then be inserted between the end of **T6** of the external memory cycle and the start of **T1** of the refresh cycle itself. The refresh address and various external strobes become active approximately one period **Tm** before **T1**. Bus signals are active until the end of **T2**, whilst **notMemRf** remains active until the end of **T6**.

For a refresh cycle, **MemnotRfD1** goes low before **notMemRf** goes low and **MemnotWrD0** goes high with the same timing as **MemnotRfD1**. All the address lines share the same timing, but only **MemAD2-11** give the refresh address. **MemAD12-30** stay high during the address period, whilst **MemAD31** remains low. Refresh cycles generate strobes **notMemS0-4** with timing as for a normal external cycle, but **notMemRd** and **notMemWrB0-3** remain high. **MemWait** operates normally during refresh cycles.

8 External memory interface

Table 8.11: Memory refresh

SYMBOL	PARAMETER	MIN	NOM	MAX	UNITS	NOTE
TRfLRfH	Refresh pulse width low	a		a+6	ns	1
TRaVS0L	Refresh address setup before notMemS0		b		ns	2
TRfLS0L	Refresh indicator setup before notMemS0		b		ns	2

Notes

1 **a** is total **Tmx**+(2 periods **Tm**).

2 **b** is total **T1**+(2 periods **Tm**) where **T1** can be from one to four periods **Tm** in length.

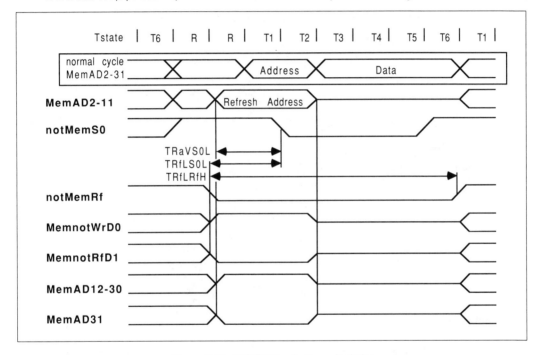

Figure 8.14: IMS T800 refresh cycle timing

8.12 MemWait

Taking **MemWait** high with the timing shown will extend the duration of **T4**. **MemWait** is sampled near to, but independent of, the falling edge of **ProcClockOut**, and should not change state in this region. By convention, **notMemS4** is used to synchronize wait state insertion. If this or another strobe is used, its delay should be such as to take the strobe low an even number of periods **Tm** after the start of **T1**, to coincide with a rising edge of **ProcClockOut**.

MemWait may be kept high indefinitely, although if dynamic memory refresh is used it should not be kept high long enough to interfere with refresh timing. **MemWait** operates normally during all cycles, including refresh and configuration cycles.

If the start of **T5** would coincide with a falling edge of **ProcClockOut** an extra wait period **Tm** (**EW**) is generated by the EMI to force coincidence with a rising edge. Rising edge coincidence is only forced if wait states are added, otherwise coincidence with a falling edge is permitted.

Table 8.12: Memory wait

SYMBOL	PARAMETER	MIN	NOM	MAX	UNITS	NOTE
TPCHWtH	Wait setup	-(a)+3			ns	1,4
TPCHWtL	Wait hold	b+3			ns	2,3,4
TWtLWtH	Delay before re-assertion of Wait	2			Tm	

Notes

1 **a** is 0.5 periods **Tm**.

2 **b** is 1.5 periods **Tm**.

3 If wait period exceeds refresh interval, refresh cycles will be lost.

4 Wait timing is independent of falling edge of **ProcClockOut**.

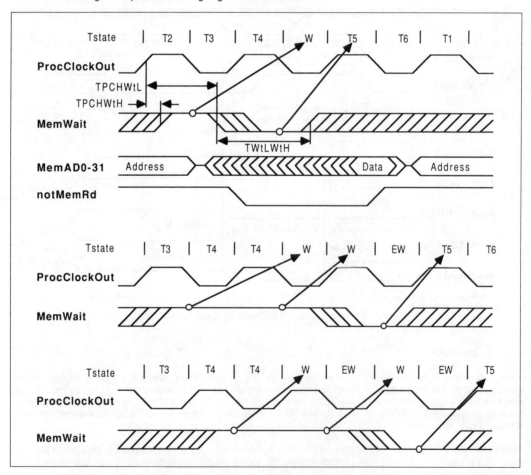

Figure 8.15: IMS T800 memory wait timing

8 External memory interface

8.13 MemReq, MemGranted

Direct memory access (DMA) can be requested at any time by taking the asynchronous **MemReq** input high. The transputer samples **MemReq** during the final period **Tm** of **T6** of both refresh and external memory cycles. To guarantee taking over the bus immediately following either, **MemReq** must be set up at least two periods **Tm** before the end of **T6**. In the absence of an external memory cycle, **MemReq** is sampled during every low period of **ProcClockOut**. The address bus is tristated two periods **Tm** after the **ProcClockOut** rising edge which follows the sample. **MemGranted** is asserted one period **Tm** after that.

Removal of **MemReq** is sampled during each low period of **ProcClockOut** and **MemGranted** is removed synchronously with the next falling edge of **ProcClockOut**. If accurate timing of DMA is required, **MemReq** should be set low coincident with a falling edge of **ProcClockOut**. Further external bus activity, either refresh, external cycles or reflection of internal cycles, will commence at the next rising edge of **ProcClockOut**.

Strobes are left in their inactive states during DMA. DMA cannot interrupt a refresh or external memory cycle, and outstanding refresh cycles will occur before the bus is released to DMA. DMA does not interfere with internal memory cycles in any way, although a program running in internal memory would have to wait for the end of DMA before accessing external memory. DMA cannot access internal memory. If DMA extends longer than one refresh interval (Memory Refresh Configuration Coding table, page 85), the DMA user becomes responsible for refresh. DMA may also inhibit an internally running program from accessing external memory.

DMA allows a bootstrap program to be loaded into external RAM ready for execution after reset. If **MemReq** is held high throughout reset, **MemGranted** will be asserted before the bootstrap sequence begins. **MemReq** must be high at least one period **TDCLDCL** of **ClockIn** before **Reset**. The circuit should be designed to ensure correct operation if **Reset** could interrupt a normal DMA cycle.

Table 8.13: Memory request

SYMBOL	PARAMETER	MIN	NOM	MAX	UNITS	NOTE
TMRHMGH	Memory request response time	4		6	Tm	1
TMRLMGL	Memory request end response time	2		4	Tm	
TADZMGH	Bus tristate before memory granted		1		Tm	
TMGLADV	Bus active after end of memory granted		1		Tm	

Notes

1 These values assume no external memory cycle is in progress. If an external cycle is active, maximum time could be (1 EMI cycle **Tmx**)+(1 refresh cycle **TRfLRfH**)+(6 periods **Tm**).

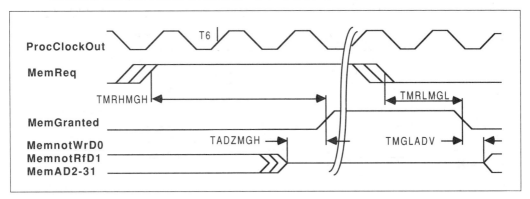

Figure 8.16: IMS T800 memory request timing

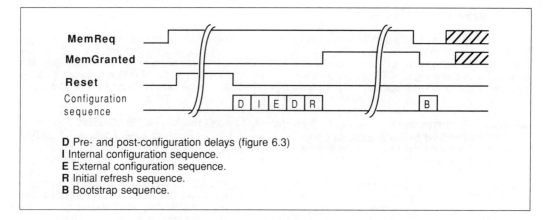

D Pre- and post-configuration delays (figure 6.3)
I Internal configuration sequence.
E External configuration sequence.
R Initial refresh sequence.
B Bootstrap sequence.

Figure 8.17: IMS T800 DMA sequence at reset

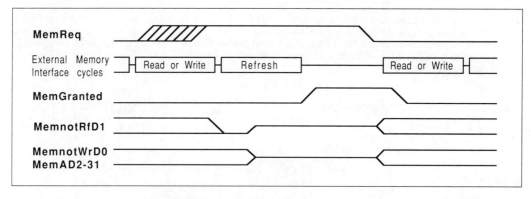

Figure 8.18: IMS T800 operation of MemReq, MemGranted with external, refresh memory cycles

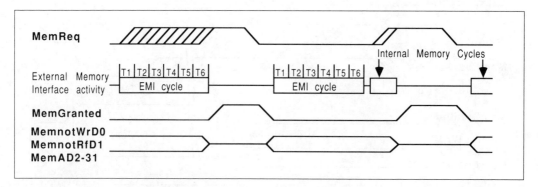

Figure 8.19: IMS T800 operation of MemReq, MemGranted with external, internal memory cycles

9 Events

EventReq and **EventAck** provide an asynchronous handshake interface between an external event and an internal process. When an external event takes **EventReq** high the external event channel (additional to the external link channels) is made ready to communicate with a process. When both the event channel and the process are ready the processor takes **EventAck** high and the process, if waiting, is scheduled. **EventAck** is removed after **EventReq** goes low.

Only one process may use the event channel at any given time. If no process requires an event to occur **EventAck** will never be taken high. Although **EventReq** triggers the channel on a transition from low to high, it must not be removed before **EventAck** is high. **EventReq** should be low during **Reset**; if not it will be ignored until it has gone low and returned high. **EventAck** is taken low when **Reset** occurs.

If the process is a high priority one and no other high priority process is running, the latency is as described on page 53. Setting a high priority task to wait for an event input is a way of interrupting a transputer program.

Table 9.1: Event

SYMBOL	PARAMETER	MIN	NOM	MAX	UNITS	NOTE
TVHKH	Event request response	0			ns	
TKHVL	Event request hold	0			ns	
TVLKL	Delay before removal of event acknowledge	0		a	ns	1
TKLVH	Delay before re-assertion of event request	0			ns	

Notes

1 **a** is **TPCLPCL** (2 periods **Tm**).

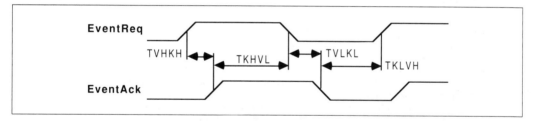

Figure 9.1: IMS T800 event timing

10 Links

Four identical INMOS bi-directional serial links provide synchronized communication between processors and with the outside world. Each link comprises an input channel and output channel. A link between two transputers is implemented by connecting a link interface on one transputer to a link interface on the other transputer Every byte of data sent on a link is acknowledged on the input of the same link, thus each signal line carries both data and control information.

The quiescent state of a link output is low. Each data byte is transmitted as a high start bit followed by a one bit followed by eight data bits followed by a low stop bit. The least significant bit of data is transmitted first. After transmitting a data byte the sender waits for the acknowledge, which consists of a high start bit followed by a zero bit. The acknowledge signifies both that a process was able to receive the acknowledged data byte and that the receiving link is able to receive another byte. The sending link reschedules the sending process only after the acknowledge for the final byte of the message has been received.

Link performance is improved over previous transputers by allowing an acknowledge packet to be sent before the data packet has been fully received. This overlapped acknowledge technique is fully compatible with all other INMOS transputer links.

The IMS T800 links support the standard INMOS communication speed of 10 Mbits per second. In addition they can be used at 5 or 20 Mbits per second. Links are not synchronised with **ClockIn** or **ProcClockOut** and are insensitive to their phases. Thus links from independently clocked systems may communicate, providing only that the clocks are nominally identical and within specification.

Links are TTL compatible and intended to be used in electrically quiet environments, between devices on a single printed circuit board or between two boards via a backplane. Direct connection may be made between devices separated by a distance of less than 300 millimetres. For longer distances a matched 100 Ohm transmission line should be used with series matching resistors **RM**. When this is done the line delay should be less than 0.4 bit time to ensure that the reflection returns before the next data bit is sent.

Buffers may be used for very long transmissions. If so, their overall propagation delay should be stable within the skew tolerance of the link, although the absolute value of the delay is immaterial.

Link speeds can be set by **LinkSpecial**, **Link0Special** and **Link123Special**. The link 0 speed can be set independently. Table 10.1 shows uni-directional and bi-directional data rates in Kbytes/second for each link speed; **LinknSpecial** is to be read as **Link0Special** when selecting link 0 speed and as **Link123Special** for the others. Data rates are quoted for a transputer using internal memory, and will be affected by a factor depending on the number of external memory accesses and the length of the external memory cycle.

Table 10.1: Speed Settings for Transputer Links

Link Special	Linkn Special	Mbits/sec	Kbytes/sec	
			Uni	Bi
0	0	10	910	1250
0	1	5	450	670
1	0	10	910	1250
1	1	20	1740	2350

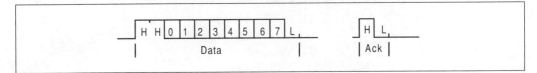

Figure 10.1: IMS T800 link data and acknowledge packets

10 Links

Table 10.2: Link

SYMBOL	PARAMETER		MIN	NOM	MAX	UNITS	NOTE
TJQr	LinkOut rise time				20	ns	
TJQf	LinkOut fall time				10	ns	
TJDr	LinkIn rise time				20	ns	
TJDf	LinkIn fall time				20	ns	
TJQJD	Buffered edge delay		0			ns	
TJBskew	Variation in TJQJD	20 Mbits/s			3	ns	1
		10 Mbits/s			10	ns	1
		5 Mbits/s			30	ns	1
CLIZ	LinkIn capacitance	@ f=1MHz			7	pF	
CLL	LinkOut load capacitance				50	pF	
RM	Series resistor for 100Ω transmission line			56		ohms	

Notes

1 This is the variation in the total delay through buffers, transmission lines, differential receivers etc., caused by such things as short term variation in supply voltages and differences in delays for rising and falling edges.

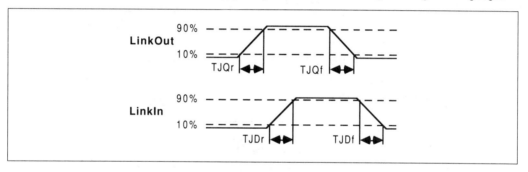

Figure 10.2: IMS T800 link timing

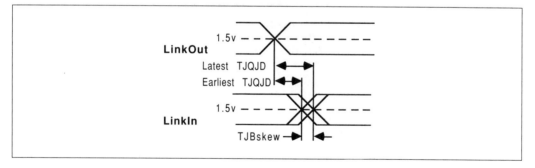

Figure 10.3: IMS T800 buffered link timing

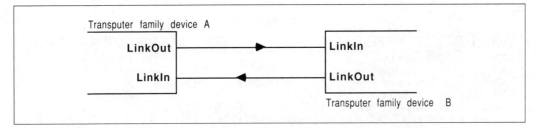

Figure 10.4: Links directly connected

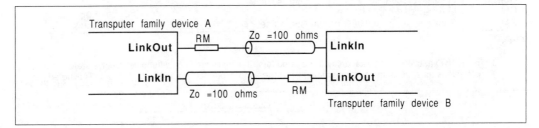

Figure 10.5: Links connected by transmission line

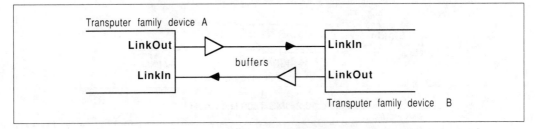

Figure 10.6: Links connected by buffers

11 Electrical specifications

11.1 DC electrical characteristics

Table 11.1: Absolute maximum ratings

SYMBOL	PARAMETER	MIN	MAX	UNITS	NOTE
VCC	DC supply voltage	0	7.0	V	1,2,3
VI, VO	Voltage on input and output pins	-0.5	VCC+0.5	V	1,2,3
II	Input current		±25	mA	4
OSCT	Output short circuit time (one pin)		1	s	2
TS	Storage temperature	-65	150	°C	2
TA	Ambient temperature under bias	-55	125	°C	2
PDmax	Maximum allowable dissipation		2	W	

Notes

1. All voltages are with respect to **GND**.

2. This is a stress rating only and functional operation of the device at these or any other conditions beyond those indicated in the operating sections of this specification is not implied. Stresses greater than those listed may cause permanent damage to the device. Exposure to absolute maximum rating conditions for extended periods may affect reliability.

3. This device contains circuitry to protect the inputs against damage caused by high static voltages or electrical fields. However, it is advised that normal precautions be taken to avoid application of any voltage higher than the absolute maximum rated voltages to this high impedance circuit. Unused inputs should be tied to an appropriate logic level such as **VCC** or **GND**.

4. The input current applies to any input or output pin and applies when the voltage on the pin is between **GND** and **VCC**.

Table 11.2: Operating conditions

SYMBOL	PARAMETER	MIN	MAX	UNITS	NOTE
VCC	DC supply voltage	4.75	5.25	V	1
VI, VO	Input or output voltage	0	VCC	V	1,2
CL	Load capacitance on any pin		50	pF	
TA	Operating temperature range	0	70	°C	3

Notes

1. All voltages are with respect to **GND**.

2. Excursions beyond the supplies are permitted but not recommended; see DC characteristics.

3. Air flow rate 400 linear ft/min transverse air flow.

Table 11.3: DC characteristics

SYMBOL	PARAMETER		MIN	MAX	UNITS	NOTE
VIH	High level input voltage		2.0	VCC+0.5	V	1,2
VIL	Low level input voltage		-0.5	0.8	V	1,2
II	Input current	@ GND<VI<VCC		±10	µA	1,2
VOH	Output high voltage	@ IOH=2mA	VCC-1		V	1,2
VOL	Output low voltage	@ IOL=4mA		0.4	V	1,2
IOS	Output short circuit current	@ GND<VO<VCC		50	mA	1,2,4
				75	mA	1,2,5
IOZ	Tristate output current	@ GND<VO<VCC		±10	µA	1,2
PD	Power dissipation			1.2	W	2,3,6
CIN	Input capacitance	@ f=1MHz		7	pF	
COZ	Output capacitance	@ f=1MHz		10	pF	

Notes

1. All voltages are with respect to **GND**.

2. Parameters measured at 4.75V<**VCC**<5.25V and 0°C<**TA**<70°C. Input clock frequency = 5MHz.

3. Power dissipation varies with output loading and program execution.

4. Current sourced from non-link outputs.

5. Current sourced from link outputs.

6. Power dissipation for processor operating at 20MHz.

11.2 Equivalent circuits

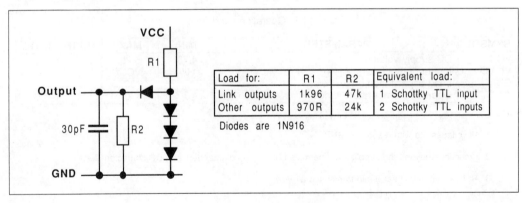

Figure 11.1: Load circuit for AC measurements

11 Electrical specifications

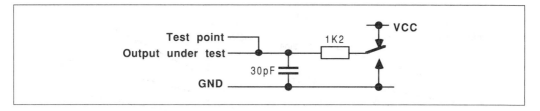

Figure 11.2: Tristate load circuit for AC measurements

11.3 AC timing characteristics

Table 11.4: Input, output edges

SYMBOL	PARAMETER	MIN	MAX	UNITS	NOTE
TDr	Input rising edges	2	20	ns	1,2
TDf	Input falling edges	2	20	ns	1,2
TQr	Output rising edges		25	ns	1
TQf	Output falling edges		15	ns	1
TS0LaHZ	Address high to tristate	a	a+6	ns	3
TS0LaLZ	Address low to tristate	a	a+6	ns	3

Notes

1 Non-link pins; see section on links.

2 All inputs except **ClockIn**; see section on **ClockIn**.

3 **a** is **T2** where **T2** can be from one to four periods **Tm** in length.
Address lines include **MemnotWrD0**, **MemnotRfD1**, **MemAD2-31**.

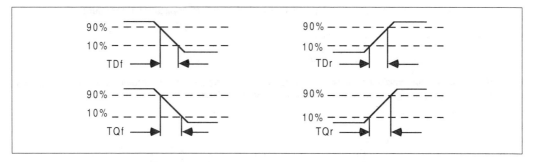

Figure 11.3: IMS T800 input and output edge timing

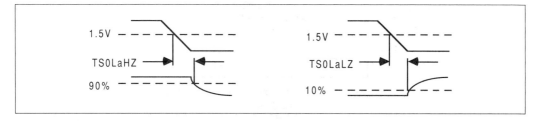

Figure 11.4: IMS T800 tristate timing relative to notMemS0

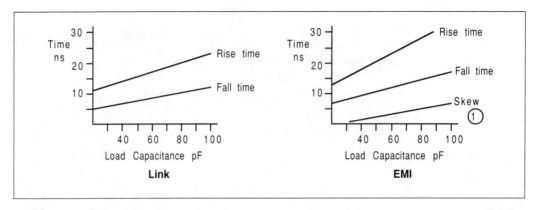

Figure 11.5: Typical rise/fall times

Notes

1. Skew is measured between **notMemS0** with a standard load (2 Schottky TTL inputs and 30pF) and **notMemS0** with a load of 2 Schottky TTL inputs and varying capacitance.

11.4 Power rating

Internal power dissipation P_{INT} of transputer and peripheral chips depends on **VCC**, as shown in figure 11.6. P_{INT} is substantially independent of temperature.

Total power dissipation P_D of the chip is

$$P_D = P_{INT} + P_{IO}$$

where P_{IO} is the power dissipation in the input and output pins; this is application dependent.

Internal working temperature T_J of the chip is

$$T_J = T_A + \theta J_A * P_D$$

where T_A is the external ambient temperature in °C and θJ_A is the junction-to-ambient thermal resistance in °C/W. θJ_A for each package is given in the Packaging Specifications section.

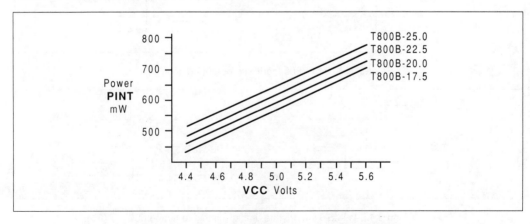

Figure 11.6: IMS T800 internal power dissipation vs VCC

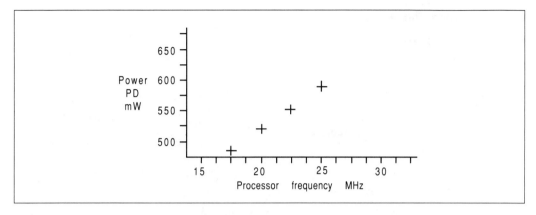

Figure 11.7: IMS T800 typical power dissipation with processor speed

12 Package specifications

12.1 84 pin grid array package

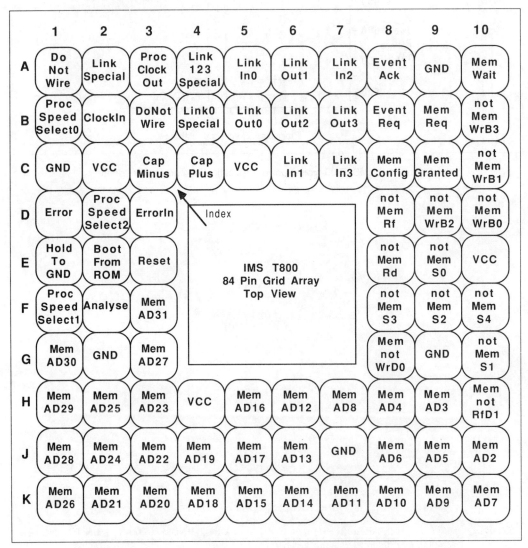

Figure 12.1: IMS T800 84 pin grid array package pinout

12 Package specifications

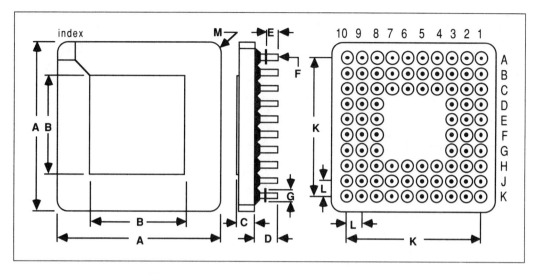

Figure 12.2: 84 pin grid array package dimensions

Table 12.1: 84 pin grid array package dimensions

DIM	Millimetres		Inches		Notes
	NOM	TOL	NOM	TOL	
A	26.924	±0.254	1.060	±0.010	
B	17.019	±0.127	0.670	±0.005	
C	2.456	±0.278	0.097	±0.011	
D	4.572	±0.127	0.180	±0.005	
E	3.302	±0.127	0.130	±0.005	
F	0.457	±0.025	0.018	±0.001	Pin diameter
G	1.143	±0.127	0.045	±0.005	Flange diameter
K	22.860	±0.127	0.900	±0.005	
L	2.540	±0.127	0.100	±0.005	
M	0.508		0.020		Chamfer

Package weight is approximately 7.2 grams

Table 12.2: 84 pin grid array package junction to ambient thermal resistance

SYMBOL	PARAMETER	MIN	NOM	MAX	UNITS	NOTE
θJA	At 400 linear ft/min transverse air flow			35	°C/W	

inmos

Chapter 4

IMS T414 engineering data

1 Introduction

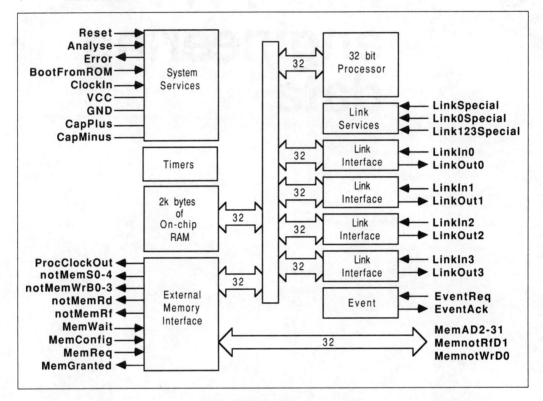

Figure 1.1: IMS T414 block diagram

1 Introduction

The IMS T414 transputer is a 32 bit CMOS microcomputer with 2 Kbytes on-chip RAM for high speed processing, a configurable memory interface and four standard INMOS communication links. The instruction set achieves efficient implementation of high level languages and provides direct support for the occam model of concurrency when using either a single transputer or a network. Procedure calls, process switching and typical interrupt latency are sub-microsecond. The IMS T414 provides high performance arithmetic and microcode support for floating point operations. A device running at 20 MHz achieves an instruction throughput of 10 MIPS.

The IMS T414 can directly access a linear address space of 4 Gbytes. The 32 bit wide memory interface uses multiplexed data and address lines and provides a data rate of up to 4 bytes every 150 nanoseconds (26.6 Mbytes/sec) for a 20 MHz device. A configurable memory controller provides all timing, control and DRAM refresh signals for a wide variety of mixed memory systems.

System Services include processor reset and bootstrap control, together with facilities for error analysis.

The INMOS communication links allow networks of transputer family products to be constructed by direct point to point connections with no external logic. The IMS T414 links support the standard operating speed of 10 Mbits per second, but also operate at 5 or 20 Mbits per second.

The IMS T414 is designed to implement the occam language, detailed in the occam Reference Manual, but also efficiently supports other languages such as C, Pascal and Fortran. Access to specific features of the IMS T414 is described in the relevant system development manual. Access to the transputer at machine level is seldom required, but if necessary refer to The Transputer Instruction Set - A Compiler Writers' Guide.

This data sheet supplies hardware implementation and characterisation details for the IMS T414. It is intended to be read in conjunction with the Transputer Architecture chapter, which details the architecture of the transputer and gives an overview of occam. For convenience of description, the IMS T414 operation is split into the basic blocks shown in figure 1.1.

2 Pin designations

Table 2.1: IMS T414 system services

Pin	In/Out	Function
VCC, GND		Power supply and return
CapPlus, CapMinus		External capacitor for internal clock power supply
ClockIn	in	Input clock
Reset	in	System reset
Error	out	Error indicator
Analyse	in	Error analysis
BootFromRom	in	Bootstrap from external ROM or from link
HoldToGND		Must be connected to **GND**
DoNotWire		Must not be wired

Table 2.2: IMS T414 external memory interface

Pin	In/Out	Function
ProcClockOut	out	Processor clock
MemnotWrD0	in/out	Multiplexed data bit 0 and write cycle warning
MemnotRfD1	in/out	Multiplexed data bit 1 and refresh warning
MemAD2-31	in/out	Multiplexed data and address bus
notMemRd	out	Read strobe
notMemWrB0-3	out	Four byte-addressing write strobes
notMemS0-4	out	Five general purpose strobes
notMemRf	out	Dynamic memory refresh indicator
MemWait	in	Memory cycle extender
MemReq	in	Direct memory access request
MemGranted	out	Direct memory access granted
MemConfig	in	Memory configuration data input

Table 2.3: IMS T414 event

Pin	In/Out	Function
EventReq	in	Event request
EventAck	out	Event request acknowledge

Table 2.4: IMS T414 link

Pin	In/Out	Function
LinkIn0-3	in	Four serial data input channels
LinkOut0-3	out	Four serial data output channels
LinkSpecial	in	Select non-standard speed as 5 or 20 Mbits/sec
Link0Special	in	Select special speed for Link 0
Link123Special	in	Select special speed for Links 1,2,3

Signal names are prefixed by **not** if they are active low, otherwise they are active high.
Pinout details for various packages are given on page 160.

3 Processor

The 32 bit processor contains instruction processing logic, instruction and work pointers, and an operand register. It directly accesses the high speed 2 Kbyte on-chip memory, which can store data or program. Where larger amounts of memory or programs in ROM are required, the processor has access to 4 Gbytes of memory via the External Memory Interface (EMI).

3.1 Registers

The design of the transputer processor exploits the availability of fast on-chip memory by having only a small number of registers; six registers are used in the execution of a sequential process. The small number of registers, together with the simplicity of the instruction set, enables the processor to have relatively simple (and fast) data-paths and control logic. The six registers are:

> The workspace pointer which points to an area of store where local variables are kept.

> The instruction pointer which points to the next instruction to be executed.

> The operand register which is used in the formation of instruction operands.

> The A, B and C registers which form an evaluation stack.

A, B and C are sources and destinations for most arithmetic and logical operations. Loading a value into the stack pushes B into C, and A into B, before loading A. Storing a value from A, pops B into A and C into B.

Expressions are evaluated on the evaluation stack, and instructions refer to the stack implicitly. For example, the *add* instruction adds the top two values in the stack and places the result on the top of the stack. The use of a stack removes the need for instructions to respecify the location of their operands. Statistics gathered from a large number of programs show that three registers provide an effective balance between code compactness and implementation complexity.

No hardware mechanism is provided to detect that more than three values have been loaded onto the stack. It is easy for the compiler to ensure that this never happens.

Any location in memory can be accessed relative to the workpointer register, enabling the workspace to be of any size.

Further register details are given in The Transputer Instruction Set - A Compiler Writers' Guide.

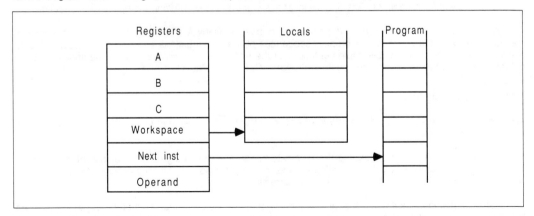

Figure 3.1: Registers

3.2 Instructions

The instruction set has been designed for simple and efficient compilation of high-level languages. All instructions have the same format, designed to give a compact representation of the operations occurring most frequently in programs.

Each instruction consists of a single byte divided into two 4-bit parts. The four most significant bits of the byte are a function code and the four least significant bits are a data value.

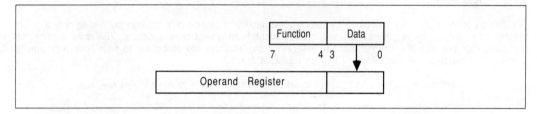

Figure 3.2: Instruction format

3.2.1 Direct functions

The representation provides for sixteen functions, each with a data value ranging from 0 to 15. Thirteen of these, shown in table 3.1, are used to encode the most important functions.

Table 3.1: Direct functions

load constant	add constant	
load local	store local	load local pointer
load non-local	store non-local	
jump	conditional jump	call

The most common operations in a program are the loading of small literal values and the loading and storing of one of a small number of variables. The *load constant* instruction enables values between 0 and 15 to be loaded with a single byte instruction. The *load local* and *store local* instructions access locations in memory relative to the workspace pointer. The first 16 locations can be accessed using a single byte instruction.

The *load non-local* and *store non-local* instructions behave similarly, except that they access locations in memory relative to the *A* register. Compact sequences of these instructions allow efficient access to data structures, and provide for simple implementations of the static links or displays used in the implementation of high level programming languages such as occam, C, Fortran, Pascal or ADA.

3.2.2 Prefix functions

Two more function codes allow the operand of any instruction to be extended in length; *prefix* and *negative prefix*.

All instructions are executed by loading the four data bits into the least significant four bits of the operand register, which is then used as the instruction's operand. All instructions except the prefix instructions end by clearing the operand register, ready for the next instruction.

The *prefix* instruction loads its four data bits into the operand register and then shifts the operand register up four places. The *negative prefix* instruction is similar, except that it complements the operand register before shifting it up. Consequently operands can be extended to any length up to the length of the operand register by a sequence of prefix instructions. In particular, operands in the range -256 to 255 can be represented using one prefix instruction.

The use of prefix instructions has certain beneficial consequences. Firstly, they are decoded and executed in the same way as every other instruction, which simplifies and speeds instruction decoding. Secondly, they simplify language compilation by providing a completely uniform way of allowing any instruction to take an operand of any size. Thirdly, they allow operands to be represented in a form independent of the processor wordlength.

3.2.3 Indirect functions

The remaining function code, *operate*, causes its operand to be interpreted as an operation on the values held in the evaluation stack. This allows up to 16 such operations to be encoded in a single byte instruction. However, the prefix instructions can be used to extend the operand of an *operate* instruction just like any other. The instruction representation therefore provides for an indefinite number of operations.

Encoding of the indirect functions is chosen so that the most frequently occurring operations are represented without the use of a prefix instruction. These include arithmetic, logical and comparison operations such as *add*, *exclusive or* and *greater than*. Less frequently occurring operations have encodings which require a single prefix operation.

3.2.4 Expression evaluation

Evaluation of expressions sometimes requires use of temporary variables in the workspace, but the number of these can be minimised by careful choice of the evaluation order.

Table 3.2: Expression evaluation

Program	Mnemonic	
x := 0	ldc	0
	stl	x
x := #24	pfix	2
	ldc	4
	stl	x
x := y + z	ldl	y
	ldl	z
	add	
	stl	x

3.2.5 Efficiency of encoding

Measurements show that about 70% of executed instructions are encoded in a single byte; that is, without the use of prefix instructions. Many of these instructions, such as *load constant* and *add* require just one processor cycle.

The instruction representation gives a more compact representation of high level language programs than more conventional instruction sets. Since a program requires less store to represent it, less of the memory bandwidth is taken up with fetching instructions. Furthermore, as memory is word accessed the processor will receive four instructions for every fetch.

Short instructions also improve the effectiveness of instruction pre-fetch, which in turn improves processor performance. There is an extra word of pre-fetch buffer, so the processor rarely has to wait for an instruction fetch before proceeding. Since the buffer is short, there is little time penalty when a jump instruction causes the buffer contents to be discarded.

3.3 Processes and concurrency

A process starts, performs a number of actions, and then either stops without completing or terminates complete. Typically, a process is a sequence of instructions. A transputer can run several processes in parallel (concurrently). Processes may be assigned either high or low priority, and there may be any number of each (page 115).

The processor has a microcoded scheduler which enables any number of concurrent processes to be executed together, sharing the processor time. This removes the need for a software kernel.

At any time, a concurrent process may be

        Active   - Being executed.
                      - On a list waiting to be executed.

        Inactive - Ready to input.
                      - Ready to output.
                      - Waiting until a specified time.

The scheduler operates in such a way that inactive processes do not consume any processor time. It allocates a portion of the processor's time to each process in turn. Active processes waiting to be executed are held in two linked lists of process workspaces, one of high priority processes and one of low priority processes (page 115). Each list is implemented using two registers, one of which points to the first process in the list, the other to the last. In the Linked Process List figure 3.3, process S is executing and P, Q and R are active, awaiting execution. Only the low priority process queue registers are shown; the high priority process ones perform in a similar manner.

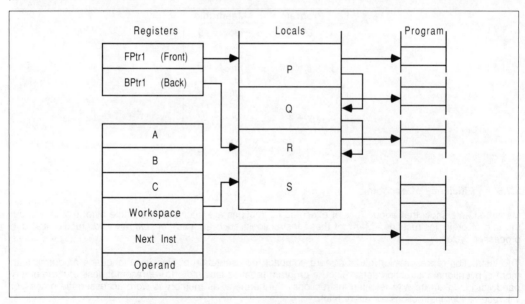

Figure 3.3: Linked process list

Table 3.3: Priority queue control registers

Function	High Priority	Low Priority
Pointer to front of active process list	*Fptr0*	*Fptr1*
Pointer to back of active process list	*Bptr0*	*Bptr1*

3 Processor

Each process runs until it has completed its action, but is descheduled whilst waiting for communication from another process or transputer, or for a time delay to complete. In order for several processes to operate in parallel, a low priority process is only permitted to run for a maximum of two time slices before it is forcibly descheduled at the next descheduling point (page 119). The time slice period is 5120 cycles of the external 5 MHz clock, giving ticks approximately 1ms apart.

A process can only be descheduled on certain instructions, known as descheduling points (page 119). As a result, an expression evaluation can be guaranteed to execute without the process being timesliced part way through.

Whenever a process is unable to proceed, its instruction pointer is saved in the process workspace and the next process taken from the list. Process scheduling pointers are updated by instructions which cause scheduling operations, and should not be altered directly. Actual process switch times are less than 1 μs, as little state needs to be saved and it is not necessary to save the evaluation stack on rescheduling.

The processor provides a number of special operations to support the process model, including *start process* and *end process*. When a main process executes a parallel construct, *start process* instructions are used to create the necessary additional concurrent processes. A *start process* instruction creates a new process by adding a new workspace to the end of the scheduling list, enabling the new concurrent process to be executed together with the ones already being executed. When a process is made active it is always added to the end of the list, and thus cannot pre-empt processes already on the same list.

The correct termination of a parallel construct is assured by use of the *end process* instruction. This uses a workspace location as a counter of the parallel construct components which have still to terminate. The counter is initialised to the number of components before the processes are *started*. Each component ends with an *end process* instruction which decrements and tests the counter. For all but the last component, the counter is non zero and the component is descheduled. For the last component, the counter is zero and the main process continues.

3.4 Priority

The IMS T414 supports two levels of priority. Priority 1 (low priority) processes are executed whenever there are no active priority 0 (high priority) processes.

High priority processes are expected to execute for a short time. If one or more high priority processes are able to proceed, then one is selected and runs until it has to wait for a communication, a timer input, or until it completes processing.

If no process at high priority is able to proceed, but one or more processes at low priority are able to proceed, then one is selected.

Low priority processes are periodically timesliced to provide an even distribution of processor time between computationally intensive tasks.

If there are **n** low priority processes, then the maximum latency from the time at which a low priority process becomes active to the time when it starts processing is 2**n**-2 timeslice periods. It is then able to execute for between one and two timeslice periods, less any time taken by high priority processes. This assumes that no process monopolises the transputer's time; i.e. it has a distribution of descheduling points (page 119).

Each timeslice period lasts for 5120 cycles of the external 5 MHz input clock (approximately 1 millisecond at the standard frequency of 5 MHz).

If a high priority process is waiting for an external channel to become ready, and if no other high priority process is active, then the interrupt latency (from when the channel becomes ready to when the process starts executing) is typically 19 processor cycles, a maximum of 58 cycles (assuming use of on-chip RAM).

3.5 Communications

Communication between processes is achieved by means of channels. Process communication is point-to-point, synchronised and unbuffered. As a result, a channel needs no process queue, no message queue and no message buffer.

A channel between two processes executing on the same transputer is implemented by a single word in memory; a channel between processes executing on different transputers is implemented by point-to-point links. The processor provides a number of operations to support message passing, the most important being *input message* and *output message*.

The *input message* and *output message* instructions use the address of the channel to determine whether the channel is internal or external. Thus the same instruction sequence can be used for both, allowing a process to be written and compiled without knowledge of where its channels are connected.

The process which first becomes ready must wait until the second one is also ready. A process performs an input or output by loading the evaluation stack with a pointer to a message, the address of a channel, and a count of the number of bytes to be transferred, and then executing an *input message* or *output message* instruction. Data is transferred if the other process is ready. If the channel is not ready or is an external one the process will deschedule.

3.6 Timers

The transputer has two 32 bit timer clocks which 'tick' periodically. The timers provide accurate process timing, allowing processes to deschedule themselves until a specific time.

One timer is accessible only to high priority processes and is incremented every microsecond, cycling completely in approximately 4295 milliseconds. The other is accessible only to low priority processes and is incremented every 64 microseconds, giving exactly 15625 ticks in one second. It has a full period of approximately 76 hours.

Table 3.4: Timer registers

Clock0	Current value of high priority (level 0) process clock
Clock1	Current value of low priority (level 1) process clock
TNextReg0	Indicates time of earliest event on high priority (level 0) timer queue
TNextReg1	Indicates time of earliest event on low priority (level 1) timer queue

The current value of the processor clock can be read by executing a *load timer* instruction. A process can arrange to perform a *timer input*, in which case it will become ready to execute after a specified time has been reached. The *timer input* instruction requires a time to be specified. If this time is in the 'past' then the instruction has no effect. If the time is in the 'future' then the process is descheduled. When the specified time is reached the process is scheduled again.

Figure 3.4 shows two processes waiting on the timer queue, one waiting for time 21, the other for time 31.

3 Processor

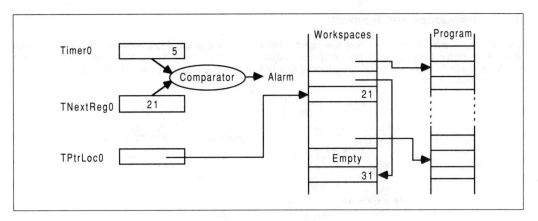

Figure 3.4: Timer registers

4   Instruction set summary

The Function Codes table 4.7. gives the basic function code set (page 112). Where the operand is less than 16, a single byte encodes the complete instruction. If the operand is greater than 15, one prefix instruction (*pfix*) is required for each additional four bits of the operand. If the operand is negative the first prefix instruction will be *nfix*.

Table 4.1: *prefix* coding

Mnemonic		Function code	Memory code
ldc	#3	#4	#43
ldc #35 **is coded as**			
pfix	#3	#2	#23
ldc	#5	#4	#45
ldc #987 **is coded as**			
pfix	#9	#2	#29
pfix	#8	#2	#28
ldc	#7	#4	#47
ldc -31 (*ldc* #FFFFFFE1) **is coded as**			
nfix	#1	#6	#61
ldc	#1	#4	#41

Tables 4.8 to 4.18 give details of the operation codes. Where an operation code is less than 16 (e.g. *add*: operation code **05**), the operation can be stored as a single byte comprising the *operate* function code **F** and the operand (**5** in the example). Where an operation code is greater than 15 (e.g. *ladd*: operation code **16**), the *prefix* function code **2** is used to extend the instruction.

Table 4.2: *operate* coding

Mnemonic		Function code	Memory code
add (op. code #5) **is coded as**			#F5
opr	add	#F	#F5
ladd (op. code #16) **is coded as**			#21F6
pfix	#1	#2	#21
opr	#6	#F	#F6

4 Instruction set summary

The Processor Cycles column refers to the number of periods **TPCLPCL** taken by an instruction executing in internal memory. The number of cycles is given for the basic operation only; where relevant the time for the *prefix* function (one cycle) should be added. For a 20 MHz transputer one cycle is 50ns. Some instruction times vary. Where a letter is included in the cycles column it is interpreted from table 4.3.

Table 4.3: Instruction set interpretation

Ident	Interpretation
b	Bit number of the highest bit set in register *A*. Bit 0 is the least significant bit.
n	Number of places shifted.
w	Number of words in the message. Part words are counted as full words. If the message is not word aligned the number of words is increased to include the part words at either end of the message.

The **DE** column of the tables indicates the descheduling/error features of an instruction as described in table 4.4.

Table 4.4: Instruction features

Ident	Feature	See page:
D	The instruction is a descheduling point	119
E	The instruction will affect the *Error* flag	119, 129

4.1 Descheduling points

The instructions in table 4.5 are the only ones at which a process may be descheduled (page 114). They are also the ones at which the processor will halt if the **Analyse** pin is asserted (page 128).

Table 4.5: Descheduling point instructions

input message	output message	output byte	output word
timer alt wait	timer input	stop on error	alt wait
jump	loop end	end process	stop process

4.2 Error instructions

The instructions in table 4.6 are the only ones which can affect the *Error* flag (page 129) directly.

Table 4.6: Error setting instructions

add	add constant	subtract	
multiply	fractional multiply	divide	remainder
long add	long subtract	long divide	
set error	testerr		cflerr
check word	check subscript from 0	check single	check count from 1

Table 4.7: IMS T414 function codes

Function Code	Memory Code	Mnemonic	Processor Cycles	Name	D E
0	0X	j	3	jump	D
1	1X	ldlp	1	load local pointer	
2	2X	pfix	1	prefix	
3	3X	ldnl	2	load non-local	
4	4X	ldc	1	load constant	
5	5X	ldnlp	1	load non-local pointer	
6	6X	nfix	1	negative prefix	
7	7X	ldl	2	load local	
8	8X	adc	1	add constant	E
9	9X	call	7	call	
A	AX	cj	2	conditional jump (not taken)	
			4	conditional jump (taken)	
B	BX	ajw	1	adjust workspace	
C	CX	eqc	2	equals constant	
D	DX	stl	1	store local	
E	EX	stnl	2	store non-local	
F	FX	opr	-	operate	

Table 4.8: IMS T414 arithmetic/logical operation codes

Operation Code	Memory Code	Mnemonic	Processor Cycles	Name	D E
46	24F6	and	1	and	
4B	24FB	or	1	or	
33	23F3	xor	1	exclusive or	
32	23F2	not	1	bitwise not	
41	24F1	shl	$n+2$	shift left	
40	24F0	shr	$n+2$	shift right	
05	F5	add	1	add	E
0C	FC	sub	1	subtract	E
53	25F3	mul	38	fractional multiply (no rounding)	E
72	27F2	fmul	35	fractional multiply (rounding)	E
			40	multiply	E
2C	22FC	div	39	divide	E
1F	21FF	rem	37	remainder	E
09	F9	gt	2	greater than	
04	F4	diff	1	difference	
52	25F2	sum	1	sum	
08	F8	prod	$b+4$	product	

4 Instruction set summary

Table 4.9: IMS T414 long arithmetic operation codes

Operation Code	Memory Code	Mnemonic	Processor Cycles	Name	DE
16	21F6	ladd	2	long add	E
38	23F8	lsub	2	long subtract	E
37	23F7	lsum	2	long sum	
4F	24FF	ldiff	2	long diff	
31	23F1	lmul	33	long multiply	
1A	21FA	ldiv	35	long divide	E
36	23F6	lshl	n+3	long shift left (n<32)	
			n-28	long shift left (n≥32)	
35	23F5	lshr	n+3	long shift right (n<32)	
			n-28	long shift right (n≥32)	
19	21F9	norm	n+5	normalise (n<32)	
			n-26	normalise (n≥32)	
			3	normalise (n=64)	

Table 4.10: IMS T414 floating point support operation codes

Operation Code	Memory Code	Mnemonic	Processor Cycles	Name	DE
73	27F3	cflerr	3	check floating point error	E
63	26F3	unpacksn	15	unpack single length fp number	
6D	26FD	roundsn	12/15	round single length fp number	
6C	26FC	postnormsn	5/30	post-normalise correction of single length fp number	
71	27F1	ldinf	1	load single length infinity	

Processor cycles are shown as **Typical/Maximum** cycles.

Table 4.11: IMS T414 general operation codes

Operation Code	Memory Code	Mnemonic	Processor Cycles	Name	DE
00	F0	rev	1	reverse	
3A	23FA	xword	4	extend to word	
56	25F6	cword	5	check word	E
1D	21FD	xdble	2	extend to double	
4C	24FC	csngl	3	check single	E
42	24F2	mint	1	minimum integer	

Table 4.12: IMS T414 indexing/array operation codes

Operation Code	Memory Code	Mnemonic	Processor Cycles	Name	D E
02	F2	bsub	1	byte subscript	
0A	FA	wsub	2	word subscript	
34	23F4	bcnt	2	byte count	
3F	23FF	wcnt	5	word count	
01	F1	lb	5	load byte	
3B	23FB	sb	4	store byte	
4A	24FA	move	2w+8	move message	

Table 4.13: IMS T414 timer handling operation codes

Operation Code	Memory Code	Mnemonic	Processor Cycles	Name	D E
22	22F2	ldtimer	2	load timer	
2B	22FB	tin	30	timer input (time future)	D
			4	timer input (time past)	D
4E	24FE	talt	4	timer alt start	
51	25F1	taltwt	15	timer alt wait (time past)	D
			48	timer alt wait (time future)	D
47	24F7	enbt	8	enable timer	
2E	22FE	dist	23	disable timer	

Table 4.14: IMS T414 input/output operation codes

Operation Code	Memory Code	Mnemonic	Processor Cycles	Name	D E
07	F7	in	2w+19	input message	D
0B	FB	out	2w+19	output message	D
0F	FF	outword	23	output word	D
0E	FE	outbyte	23	output byte	D
12	21F2	resetch	3	reset channel	
43	24F3	alt	2	alt start	
44	24F4	altwt	5	alt wait (channel ready)	D
			17	alt wait (channel not ready)	D
45	24F5	altend	4	alt end	
49	24F9	enbs	3	enable skip	
30	23F0	diss	4	disable skip	
48	24F8	enbc	7	enable channel (ready)	
			5	enable channel (not ready)	
2F	22FF	disc	8	disable channel	

4 Instruction set summary

Table 4.15: IMS T414 control operation codes

Operation Code	Memory Code	Mnemonic	Processor Cycles	Name	D E
20	22F0	ret	5	return	
1B	21FB	ldpi	2	load pointer to instruction	
3C	23FC	gajw	2	general adjust workspace	
06	F6	gcall	4	general call	
21	22F1	lend	10	loop end (loop)	D
			5	loop end (exit)	D

Table 4.16: IMS T414 scheduling operation codes

Operation Code	Memory Code	Mnemonic	Processor Cycles	Name	D E
0D	FD	startp	12	start process	D
03	F3	endp	13	end process	D
39	23F9	runp	10	run process	
15	21F5	stopp	11	stop process	
1E	21FE	ldpri	1	load current priority	

Table 4.17: IMS T414 error handling operation codes

Operation Code	Memory Code	Mnemonic	Processor Cycles	Name	D E
13	21F3	csub0	2	check subscript from 0	E
4D	24FD	ccnt1	3	check count from 1	E
29	22F9	testerr	2	test error false and clear (no error)	
			3	test error false and clear (error)	
10	21F0	seterr	1	set error	E
55	25F5	stoperr	2	stop on error (no error)	D
57	25F7	clrhalterr	1	clear halt-on-error	
58	25F8	sethalterr	1	set halt-on-error	
59	25F9	testhalterr	2	test halt-on-error	

Table 4.18: IMS T414 processor initialisation operation codes

Operation Code	Memory Code	Mnemonic	Processor Cycles	Name	D E
2A	22FA	testpranal	2	test processor analysing	
3E	23FE	saveh	4	save high priority queue registers	
3D	23FD	savel	4	save low priority queue registers	
18	21F8	sthf	1	store high priority front pointer	
50	25F0	sthb	1	store high priority back pointer	
1C	21FC	stlf	1	store low priority front pointer	
17	21F7	stlb	1	store low priority back pointer	
54	25F4	sttimer	1	store timer	

5 System services

System services include all the necessary logic to initialise and sustain operation of the device. They also include error handling and analysis facilities.

5.1 Power

Power is supplied to the device via the **VCC** and **GND** pins. Several of each are provided to minimise inductance within the package. All supply pins must be connected. The supply must be decoupled close to the chip by at least one 100nF low inductance (e.g. ceramic) capacitor between **VCC** and **GND**. Four layer boards are recommended; if two layer boards are used, extra care should be taken in decoupling.

Input voltages must not exceed specification with respect to **VCC** and **GND**, even during power-up and power-down ramping, otherwise *latchup* can occur. CMOS devices can be permanently damaged by excessive periods of latchup.

5.2 CapPlus, CapMinus

The internally derived power supply for internal clocks requires an external low leakage, low inductance $1\mu F$ capacitor to be connected between **CapPlus** and **CapMinus**. A ceramic capacitor is preferred, with an impedance less than 3 ohms between 100 KHz and 20 MHz. If a polarised capacitor is used the negative terminal should be connected to **CapMinus**. Total PCB track length should be less than 50mm. The connections must not touch power supplies or other noise sources.

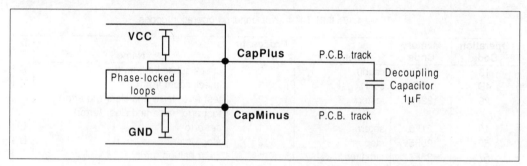

Figure 5.1: Recommended PLL decoupling

5.3 ClockIn

Transputer family components use a standard clock frequency, supplied by the user on the **ClockIn** input. The nominal frequency of this clock for all transputer family components is 5MHz, regardless of device type, transputer word length or processor cycle time. High frequency internal clocks are derived from **ClockIn**, simplifying system design and avoiding problems of distributing high speed clocks externally.

A number of transputer devices may be connected to a common clock, or may have individual clocks providing each one meets the specified stability criteria. In a multi-clock system the relative phasing of **ClockIn** clocks is not important, due to the asynchronous nature of the links. Mark/space ratio is unimportant provided the specified limits of **ClockIn** pulse widths are met.

Oscillator stability is important. **ClockIn** must be derived from a crystal oscillator; RC oscillators are not sufficiently stable. **ClockIn** must not be distributed through a long chain of buffers. Clock edges must be monotonic and remain within the specified voltage and time limits.

5 System services

Table 5.1: Input clock

SYMBOL	PARAMETER	MIN	NOM	MAX	UNITS	NOTE
TDCLDCH	ClockIn pulse width low	40			ns	
TDCHDCL	ClockIn pulse width high	40			ns	
TDCLDCL	ClockIn period		200		ns	1,3
TDCerror	ClockIn timing error			±0.5	ns	2
TDC1DC2	Difference in ClockIn for 2 linked devices			400	ppm	3
TDCr	ClockIn rise time			10	ns	4
TDCf	ClockIn fall time			8	ns	4

Notes

1 Measured between corresponding points on consecutive falling edges.

2 Variation of individual falling edges from their nominal times.

3 This value allows the use of 200ppm crystal oscillators for two devices connected together by a link.

4 Clock transitions must be monotonic within the range **VIH** to **VIL** (page 156).

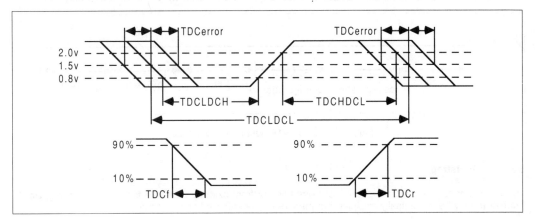

Figure 5.2: ClockIn timing

5.4 Reset

Reset can go high with **VCC**, but must at no time exceed the maximum specified voltage for **VIH**. After **VCC** is valid **ClockIn** should be running for a minimum period **TDCVRL** before the end of **Reset**. The falling edge of **Reset** initialises the transputer, triggers the memory configuration sequence and starts the bootstrap routine. Link outputs are forced low during reset; link inputs and **EventReq** should be held low. Memory request (DMA) must not occur whilst **Reset** is high but can occur before bootstrap (page 150).

After the end of **Reset** there will be a delay of 144 periods of **ClockIn** (figure 5.3). Following this, the **MemWrD0**, **MemRfD1** and **MemAD2-31** pins will be scanned to check for the existence of a pre-programmed memory interface configuration (page 141). This lasts for a further 144 periods of **ClockIn**. Regardless of whether a configuration was found, 36 configuration read cycles will then be performed on external memory using the default memory configuration (page 142), in an attempt to access the external configuration ROM. A delay will then occur, its period depending on the actual configuration. Finally eight complete and consecutive refresh cycles will initialise any dynamic RAM, using the new memory configuration. If the memory configuration does not enable refresh of dynamic RAM the refresh cycles will be replaced by an equivalent delay with no external memory activity.

If **BootFromRom** is high bootstrapping will then take place immediately, using data from external memory; otherwise the transputer will await an input from any link. The processor will be in the low priority state.

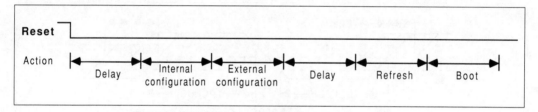

Figure 5.3: IMS T414 post-reset sequence

5.5 Bootstrap

The transputer can be bootstrapped either from a link or from external ROM. To facilitate debugging, **BootFromRom** may be dynamically changed but must obey the specified timing restrictions.

If **BootFromRom** is connected high (e.g. to **VCC**) the transputer starts to execute code from the top two bytes in external memory, at address #7FFFFFFE. This location should contain a backward jump to a program in ROM. The processor is in the low priority state, and the *W* register points to *MemStart* (page 130).

If **BootFromRom** is connected low (e.g. to **GND**) the transputer will wait for the first bootstrap message to arrive on any one of its links. The transputer is ready to receive the first byte on a link within two processor cycles **TPCLPCL** after **Reset** goes low.

If the first byte received (the control byte) is greater than 1 it is taken as the quantity of bytes to be input. The following bytes, to that quantity, are then placed in internal memory starting at location *MemStart*. Following reception of the last byte the transputer will start executing code at *MemStart* as a low priority process. The memory space immediately above the loaded code is used as work space. Messages arriving on other links after the control byte has been received and on the bootstrapping link after the last bootstrap byte will be retained until a process inputs from them.

5 System services

Table 5.2: Reset and Analyse

SYMBOL	PARAMETER	MIN	NOM	MAX	UNITS	NOTE
TPVRH	Power valid before Reset	10			ms	
TRHRL	Reset pulse width high	8			ClockIn	1
TDCVRL	ClockIn running before Reset end	10			ms	2
TAHRH	Analyse setup before Reset	3			ms	
TRLAL	Analyse hold after Reset end	1			ns	
TBRVRL	BootFromRom setup	0			ms	
TRLBRX	BootFromRom hold after Reset	50			ms	
TALBRX	BootFromRom hold after Analyse	50			ms	

Notes

1 Full periods of **ClockIn TDCLDCL** required.

2 At power-on reset.

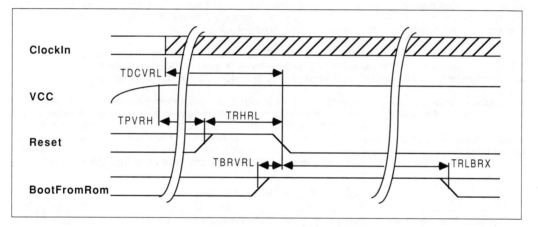

Figure 5.4: Transputer reset timing with Analyse low

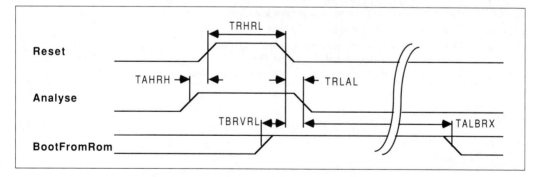

Figure 5.5: Transputer reset and analyse timing

5.6 Peek and poke

Any location in internal or external memory can be interrogated and altered when the transputer is waiting for a bootstrap from link. If the control byte is 0 then eight more bytes are expected on the same link. The first four byte word is taken as an internal or external memory address at which to poke (write) the second four byte word. If the control byte is 1 the next four bytes are used as the address from which to peek (read) a word of data; the word is sent down the output channel of the same link.

Following such a peek or poke, the transputer returns to its previously held state. Any number of accesses may be made in this way until the control byte is greater than 1, when the transputer will commence reading its bootstrap program. Any link can be used, but addresses and data must be transmitted via the same link as the control byte.

5.7 Analyse

If **Analyse** is taken high when the transputer is running, the transputer will halt at the next descheduling point (page 119). From **Analyse** being asserted, the processor will halt within three time slice periods plus the time taken for any high priority process to complete. As much of the transputer status is maintained as is necessary to permit analysis of the halted machine. Memory refresh continues.

Input links will continue with outstanding transfers. Output links will not make another access to memory for data but will transmit only those bytes already in the link buffer. Providing there is no delay in link acknowledgement, the links should be inactive within a few microseconds of the transputer halting.

Reset should not be asserted before the transputer has halted and link transfers have ceased. When **Reset** is taken low whilst **Analyse** is high, neither the memory configuration sequence nor the block of eight refresh cycles will occur; the previous memory configuration will be used for any external memory accesses. If **BootFromRom** is high the transputer will bootstrap as soon as **Analyse** is taken low, otherwise it will await a control byte on any link. If **Analyse** is taken low without **Reset** going high the transputer state and operation are undefined. After the end of a valid **Analyse** sequence the registers have the values given in table 5.3.

Table 5.3: Register values after analyse

I		*MemStart* if bootstrapping from a link, or the external memory bootstrap address if bootstrapping from ROM.
W		*MemStart* if bootstrapping from ROM, or the address of the first free word after the bootstrap program if bootstrapping from link.
A		The value of *I* when the processor halted.
B		The value of *W* when the processor halted, together with the priority of the process when the transputer was halted (i.e. the *W* descriptor).
C		The ID of the bootstrapping link if bootstrapping from link.

5.8 Error

The **Error** pin is connected directly to the internal *Error* flag and follows the state of that flag. If **Error** is high it indicates an error in one of the processes caused, for example, by arithmetic overflow, divide by zero, array bounds violation or software setting the flag directly (page 119). Once set, the *Error* flag is only cleared by executing the instruction *testerr* (page 118). The error is not cleared by processor reset, in order that analysis can identify any errant transputer (page 128).

A process can be programmed to stop if the *Error* flag is set; it cannot then transmit erroneous data to other processes, but processes which do not require that data can still be scheduled. Eventually all processes which rely, directly or indirectly, on data from the process in error will stop through lack of data.

By setting the *HaltOnError* flag the transputer itself can be programmed to halt if *Error* becomes set. If *Error* becomes set after *HaltOnError* has been set, all processes on that transputer will cease but will not necessarily cause other transputers in a network to halt. Setting *HaltOnError* after *Error* will not cause the transputer to halt; this allows the processor reset and analyse facilities to function with the flags in indeterminate states.

An alternative method of error handling is to have the errant process or transputer cause all transputers to halt. This can be done by applying the **Error** output signal of the errant transputer to the **EventReq** pin of a suitably programmed master transputer. Since the process state is preserved when stopped by an error, the master transputer can then use the analyse function to debug the fault. When using such a circuit, note that the *Error* flag is in an indeterminate state on power up; the circuit and software should be designed with this in mind.

Error checks can be removed completely to optimise the performance of a proven program; any unexpected error then occurring will have an arbitrary undefined effect.

If a high priority process pre-empts a low priority one, status of the *Error* and *HaltOnError* flags is saved for the duration of the high priority process and restored at the conclusion of it. Status of the *Error* flag is transmitted to the high priority process but the *HaltOnError* flag is cleared before the process starts. Either flag can be altered in the process without upsetting the error status of any complex operation being carried out by the pre-empted low priority process.

In the event of a transputer halting because of *HaltOnError*, the links will finish outstanding transfers before shutting down. If **Analyse** is asserted then all inputs continue but outputs will not make another access to memory for data. Memory refresh will continue to take place.

After halting due to the *Error* flag changing from 0 to 1 whilst *HaltOnError* is set, register *I* points two bytes past the instruction which set *Error*. After halting due to the **Analyse** pin being taken high, register *I* points one byte past the instruction being executed. In both cases *I* will be copied to register *A*.

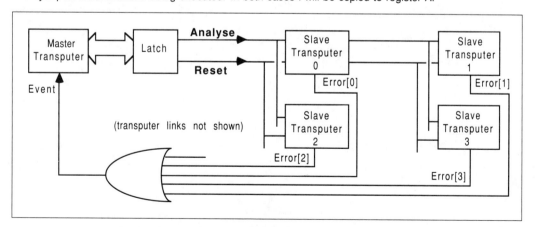

Figure 5.6: Error handling in a multi-transputer system

6 Memory

The IMS T414 has 2 Kbytes of fast internal static memory for high rates of data throughput. Each internal memory access takes one processor cycle **ProcClockOut** (page 132). The transputer can also access 4 Gbytes of external memory space. Internal and external memory are part of the same linear address space.

IMS T414 memory is byte addressed, with words aligned on four-byte boundaries. The least significant byte of a word is the lowest addressed byte.

The bits in a byte are numbered 0 to 7, with bit 0 the least significant. The bytes are numbered from 0, with byte 0 the least significant. In general, wherever a value is treated as a number of component values, the components are numbered in order of increasing numerical significance, with the least significant component numbered 0. Where values are stored in memory, the least significant component value is stored at the lowest (most negative) address.

Internal memory starts at the most negative address #80000000 and extends to #800007FF. User memory begins at #80000048; this location is given the name *MemStart*.

A reserved area at the bottom of internal memory is used to implement link and event channels.

Two words of memory are reserved for timer use, *TPtrLoc0* for high priority processes and *TPtrLoc1* for low priority processes. They either indicate the relevant priority timer is not in use or point to the first process on the timer queue at that priority level.

Values of certain processor registers for the current low priority process are saved in the reserved *IntSaveLoc* locations when a high priority process pre-empts a low priority one.

External memory space starts at #80000800 and extends up through #00000000 to #7FFFFFFF. Memory configuration data and ROM bootstrapping code must be in the most positive address space, starting at #7FFFFF6C and #7FFFFFFE respectively. Address space immediately below this is conventionally used for ROM based code.

6 Memory

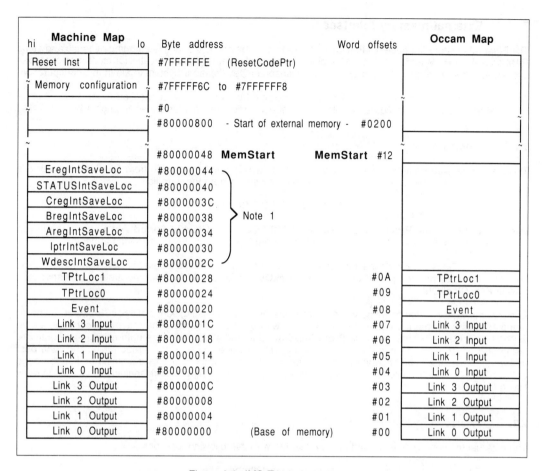

Figure 6.1: IMS T414 memory map

These locations are used as auxiliary processor registers and should not be manipulated by the user. Like processor registers, their contents may be useful for implementing debugging tools (**Analyse**, page 128). For details see The Transputer Instruction Set - A Compiler Writers' Guide.

7 External memory interface

The External Memory Interface (EMI) allows access to a 32 bit address space, supporting dynamic and static RAM as well as ROM and EPROM. EMI timing can be configured at **Reset** to cater for most memory types and speeds, and a program is supplied with the Transputer Development System to aid in this configuration.

There are 13 internal configurations which can be selected by a single pin connection (page 141). If none are suitable the user can configure the interface to specific requirements, as shown in page 142.

7.1 ProcClockOut

This clock is derived from the internal processor clock, which is in turn derived from **ClockIn**. Its period is equal to one internal microcode cycle time, and can be derived from the formula

$$TPCLPCL = TDCLDCL / PLLx$$

where **TPCLPCL** is the **ProcClockOut Period**, **TDCLDCL** is the **ClockIn Period** and **PLLx** is the phase lock loop factor for the relevant speed part, obtained from the ordering details (Ordering appendix).

The time value **Tm** is used to define the duration of **Tstates** and, hence, the length of external memory cycles; its value is exactly half the period of one **ProcClockOut** cycle (0.5∗**TPCLPCL**), regardless of mark/space ratio of **ProcClockOut**.

Edges of the various external memory strobes coincide with rising or falling edges of **ProcClockOut**. It should be noted, however, that there is a skew associated with each coincidence. The value of skew depends on whether coincidence occurs when the **ProcClockOut** edge and strobe edge are both rising, when both are falling or if either is rising when the other is falling. Timing values given in the strobe tables show the best and worst cases. If a more accurate timing relationship is required, the exact **Tstate** timing and strobe edge to **ProcClockOut** relationships should be calculated and the correct skew factors applied from the edge skew timing table 7.4.

7.2 Tstates

The external memory cycle is divided into six **Tstates** with the following functions:

- **T1** Address setup time before address valid strobe.
- **T2** Address hold time after address valid strobe.
- **T3** Read cycle tristate or write cycle data setup.
- **T4** Extendable data setup time.
- **T5** Read or write data.
- **T6** Data hold.

Under normal conditions each **Tstate** may be from one to four periods **Tm** long, the duration being set during memory configuration. The default condition on **Reset** is that all **Tstates** are the maximum four periods **Tm** long to allow external initialisation cycles to read slow ROM.

Period **T4** can be extended indefinitely by adding externally generated wait states.

An external memory cycle is always an even number of periods **Tm** in length and the start of **T1** always coincides with a rising edge of **ProcClockOut**. If the total configured quantity of periods **Tm** is an odd number, one extra period **Tm** will be added at the end of **T6** to force the start of the next **T1** to coincide with a rising edge of **ProcClockOut**. This period is designated **E** in configuration diagrams (page 142).

7 External memory interface

Table 7.1: ProcClockOut

SYMBOL	PARAMETER	MIN	NOM	MAX	UNITS	NOTE
TPCLPCL	ProcClockOut period	a-1	a	a+1	ns	1
TPCHPCL	ProcClockOut pulse width high	b-2.5	b	b+2.5	ns	2
TPCLPCH	ProcClockOut pulse width low		c		ns	3
Tm	ProcClockOut half cycle	b-0.5	b	b+0.5	ns	2
TPCstab	ProcClockOut stability			4	%	4

Notes

1 **a** is **TDCLDCL/PLLx**.

2 **b** is 0.5∗**TPCLPCL** (half the processor clock period).

3 **c** is **TPCLPCL-TPCHPCL**.

4 Stability is the variation of cycle periods between two consecutive cycles, measured at corresponding points on the cycles.

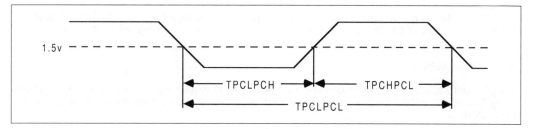

Figure 7.1: IMS T414 ProcClockOut timing

7.3 Internal access

During an internal memory access cycle the external memory interface bus **MemAD2-31** reflects the word address used to access internal RAM, **MemnotWrD0** reflects the read/write operation and **MemnotRfD1** is high; all control strobes are inactive. This is true unless and until a memory refresh cycle or DMA (memory request) activity takes place, when the bus will carry the appropriate external address or data.

The bus activity is not adequate to trace the internal operation of the transputer in full, but may be used for hardware debugging in conjuction with peek and poke (page 128).

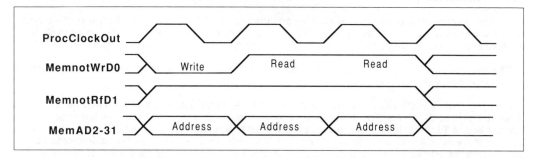

Figure 7.2: IMS T414 bus activity for internal memory cycle

7.4 MemAD2-31

External memory addresses and data are multiplexed on one bus. Only the top 30 bits of address are output on the external memory interface, using pins **MemAD2-31**. They are normally output only during **Tstates T1** and **T2**, and should be latched during this time. Byte addressing is carried out internally by the transputer for read cycles. For write cycles the relevant bytes in memory are addressed by the write strobes **notMemWrB0-3**.

The data bus is 32 bits wide. It uses **MemAD2-31** for the top 30 bits and **MemnotRfD1** and **MemnotWrD0** for the lower two bits. Read cycle data may be set up on the bus at any time after the start of **T3**, but must be valid when the transputer reads it at the end of **T5**. Data may be removed any time during **T6**, but must be off the bus no later than the end of that period.

Write data is placed on the bus at the start of **T3** and removed at the end of **T6**. If **T6** is extended to force the next cycle **Tmx** (page 134) to start on a rising edge of **ProcClockOut**, data will be valid during this time also.

7.5 MemnotWrD0

During **T1** and **T2** this pin will be low if the cycle is a write cycle, otherwise it will be high. During **Tstates T3** to **T6** it becomes bit 0 of the data bus. In both cases it follows the general timing of **MemAD2-31**.

7.6 MemnotRfD1

During **T1** and **T2**, this pin is low if the address on **MemAD2-31** is a refresh address, otherwise it is high. During **Tstates T3** to **T6** it becomes bit 1 of the data bus. In both cases it follows the general timing of **MemAD2-31**.

7.7 notMemRd

For a read cycle the read strobe **notMemRd** is low during **T4** and **T5**. Data is read by the transputer on the rising edge of this strobe, and may be removed immediately afterwards. If the strobe duration is insufficient it may be extended by adding extra periods **Tm** to either or both of the **Tstates T4** and **T5**. Further extension may be obtained by inserting wait states at the end of **T4**.

In the read cycle timing diagrams **ProcClockOut** is included as a guide only; it is shown with each **Tstate** configured to one period **Tm**.

7.8 notMemS0-4

To facilitate control of different types of memory and devices, the EMI is provided with five strobe outputs, four of which can be configured by the user. The strobes are conventionally assigned the functions shown in the read and write cycle diagrams, although there is no compulsion to retain these designations.

notMemS0 is a fixed format strobe. Its leading edge is always coincident with the start of **T2** and its trailing edge always coincident with the end of **T5**.

The leading edge of **notMemS1** is always coincident with the start of **T2**, but its duration may be configured to be from zero to 31 periods **Tm**. Regardless of the configured duration, the strobe will terminate no later than the end of **T6**. The strobe is sometimes programmed to extend beyond the normal end of **Tmx**. When wait states are inserted into an EMI cycle the end of **Tmx** is delayed, but the potential active duration of the strobe is not altered. Thus the strobe can be configured to terminate relatively early under certain conditions (page 148). If **notMemS1** is configured to be zero it will never go low.

notMemS2, **notMemS3** and **notMemS4** are identical in operation. They all terminate at the end of **T5**, but the start of each can be delayed from one to 31 periods **Tm** beyond the start of **T2**. If the duration of one of

7 External memory interface

these strobes would take it past the end of **T5** it will stay high. This can be used to cause a strobe to become active only when wait states are inserted. If one of these strobes is configured to zero it will never go high. Figure 7.5 shows the effect of **Wait** on strobes in more detail; each division on the scale is one period **Tm**.

Table 7.2: Read

SYMBOL	PARAMETER	MIN	NOM	MAX	UNITS	NOTE
TaZdV	Address tristate to data valid	0			ns	
TdVRdH	Data setup before read	20			ns	
TRdHdX	Data hold after read	0			ns	
TS0LRdL	notMemS0 before start of read	a-2	a	a+2	ns	1
TS0HRdH	End of read from end of notMemS0	-1		1	ns	
TRdLRdH	Read period	b		b+6	ns	2

Notes

1 **a** is total of **T2**+**T3** where **T2**, **T3** can be from one to four periods **Tm** each in length.

2 **b** is total of **T4**+**Twait**+**T5** where **T4**, **T5** can be from one to four periods **Tm** each in length and **Twait** may be any number of periods **Tm** in length.

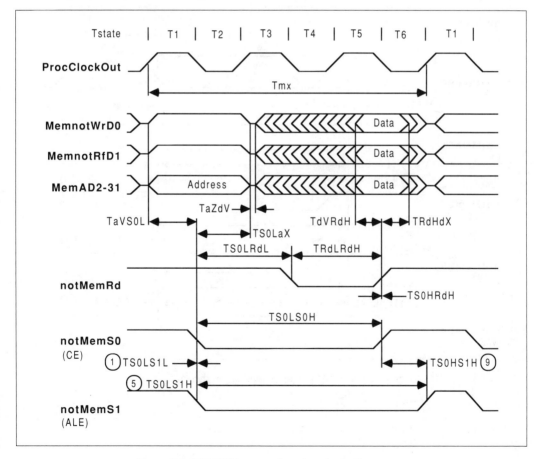

Figure 7.3: IMS T414 external read cycle: static memory

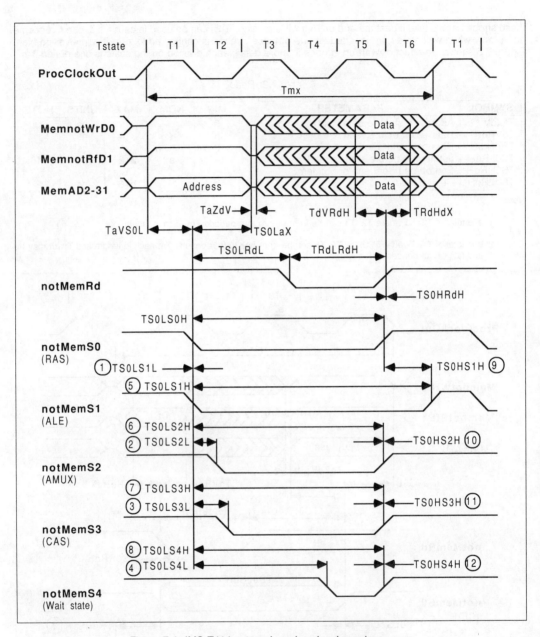

Figure 7.4: IMS T414 external read cycle: dynamic memory

7 External memory interface

Table 7.3: IMS T414 strobe timing

SYMBOL	(n)	PARAMETER	MIN	NOM	MAX	UNITS	NOTE
TaVS0L		Address setup before notMemS0		a		ns	1
TS0LaX		Address hold after notMemS0		b		ns	2
TS0LS0H		notMemS0 pulse width low	c		c+6	ns	3
TS0LS1L	1	notMemS1 from notMemS0	0		2	ns	
TS0LS1H	5	notMemS1 end from notMemS0	d		d+6	ns	4,6
TS0HS1H	9	notMemS1 end from notMemS0 end	e-1		e+4	ns	5,6
TS0LS2L	2	notMemS2 delayed after notMemS0	f-1		f+4	ns	7
TS0LS2H	6	notMemS2 end from notMemS0	c+4		c+8	ns	3
TS0HS2H	10	notMemS2 end from notMemS0 end	0		2	ns	
TS0LS3L	3	notMemS3 delayed after notMemS0	f-1		f+3	ns	7
TS0LS3H	7	notMemS3 end from notMemS0	c+4		c+8	ns	3
TS0HS3H	11	notMemS3 end from notMemS0 end	0		2	ns	
TS0LS4L	4	notMemS4 delayed after notMemS0	f-1		f+2	ns	7
TS0LS4H	8	notMemS4 end from notMemS0	c+4		c+8	ns	3
TS0HS4H	12	notMemS4 end from notMemS0 end	0		2	ns	
Tmx		Complete external memory cycle		g			8

Notes

1. **a** is **T1** where **T1** can be from one to four periods **Tm** in length.

2. **b** is **T2** where **T2** can be from one to four periods **Tm** in length.

3. **c** is total of **T2+T3+T4+Twait+T5** where **T2**, **T3**, **T4**, **T5** can be from one to four periods **Tm** each in length and **Twait** may be any number of periods **Tm** in length.

4. **d** can be from zero to 31 periods **Tm** in length.

5. **e** can be from -27 to +4 periods **Tm** in length.

6. If the configuration would cause the strobe to remain active past the end of **T6** it will go high at the end of **T6**. If the strobe is configured to zero periods **Tm** it will remain high throughout the complete cycle **Tmx**.

7. **f** can be from zero to 31 periods **Tm** in length. If this length would cause the strobe to remain active past the end of **T5** it will go high at the end of **T5**. If the strobe value is zero periods **Tm** it will remain low throughout the complete cycle **Tmx**.

8. **g** is one complete external memory cycle comprising the total of **T1+T2+T3+T4+Twait+T5+T6** where **T1**, **T2**, **T3**, **T4**, **T5** can be from one to four periods **Tm** each in length, **T6** can be from one to five periods **Tm** in length and **Twait** may be zero or any number of periods **Tm** in length.

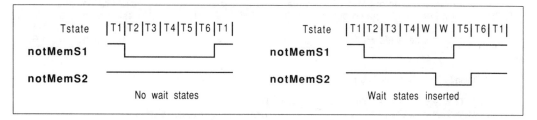

Figure 7.5: IMS T414 effect of wait states on strobes

Table 7.4: Strobe S0 to ProcClockOut skew

SYMBOL	PARAMETER	MIN	NOM	MAX	UNITS	NOTE
TPCHS0H	Strobe rising from ProcClockOut rising	0		3	ns	
TPCLS0H	Strobe rising from ProcClockOut falling	1		4	ns	
TPCHS0L	Strobe falling from ProcClockOut rising	-3		0	ns	
TPCLS0L	Strobe falling from ProcClockOut falling	-1		2	ns	

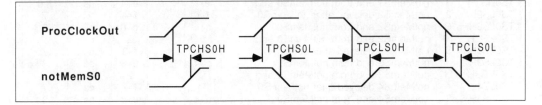

Figure 7.6: IMS T414 skew of notMemS0 to ProcClockOut

7.9 notMemWrB0-3

Because the transputer uses word addressing, four write strobes are provided; one to write each byte of the word. **notMemWrB0** addresses the least significant byte.

The transputer has both early and late write cycle modes. For a late write cycle the relevant write strobes **notMemWrB0-3** are low during **T4** and **T5**; for an early write they are also low during **T3**. Data should be latched into memory on the rising edge of the strobes in both cases, although it is valid until the end of **T6**. If the strobe duration is insufficient, it may be extended at configuration time by adding extra periods **Tm** to either or both of **Tstates T4** and **T5** for both early and late modes. For an early cycle they may also be added to **T3**. Further extension may be obtained by inserting wait states at the end of **T4**. If the data hold time is insufficient, extra periods **Tm** may be added to **T6** to extend it.

Table 7.5: Write

SYMBOL	PARAMETER	MIN	NOM	MAX	UNITS	NOTE
TdVWrH	Data setup before write	d			ns	1,5
TWrHdX	Data hold after write	a			ns	1,2
TS0LWrL	notMemS0 before start of early write	b-3		b+2	ns	1,3
	notMemS0 before start of late write	c-3		c+2	ns	1,4
TS0HWrH	End of write from end of notMemS0	-2		2	ns	1
TWrLWrH	Early write pulse width	d		d+6	ns	1,5
	Late write pulse width	e		e+6	ns	1,6

Notes

1 Timing is for all write strobes **notMemWrB0-3**.

2 **a** is **T6** where **T6** can be from one to five periods **Tm** in length.

3 **b** is **T2** where **T2** can be from one to four periods **Tm** in length.

4 **c** is total of **T2**+**T3** where **T2**, **T3** can be from one to four periods **Tm** each in length.

5 **d** is total of **T3**+**T4**+**Twait**+**T5** where **T3**, **T4**, **T5** can be from one to four periods **Tm** each in length and **Twait** may be zero or any number of periods **Tm** in length.

6 **e** is total of **T4**+**Twait**+**T5** where **T4**, **T5** can be from one to four periods **Tm** each in length and **Twait** may be zero or any number of periods **Tm** in length.

7 External memory interface

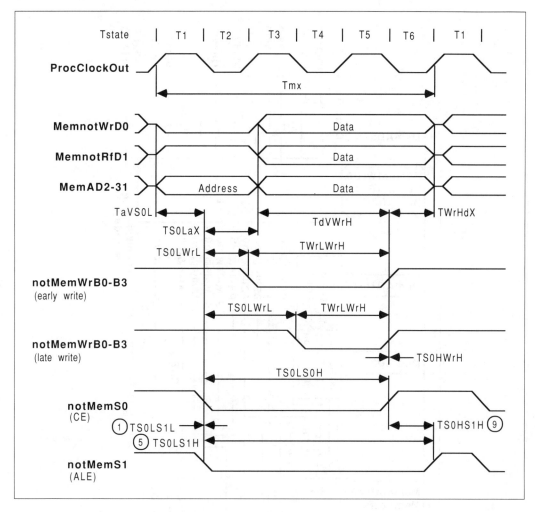

Figure 7.7: IMS T414 external write cycle

In the write cycle timing diagram **ProcClockOut** is included as a guide only; it is shown with each **Tstate** configured to one period **Tm**. The strobe is inactive during internal memory cycles.

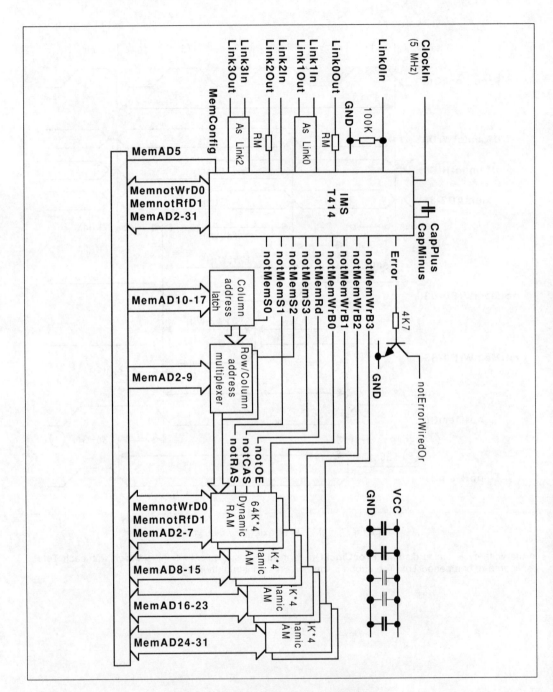

Figure 7.8: IMS T414 application

7 External memory interface

7.10 MemConfig

MemConfig is an input pin used to read configuration data when setting external memory interface (EMI) characteristics. It is read by the processor on two occasions after **Reset** goes low; first to check if one of the preset internal configurations is required, then to determine a possible external configuration.

7.10.1 Internal configuration

The internal configuration scan comprises 64 periods **TDCLDCL** of **ClockIn** during the internal scan period of 144 **ClockIn** periods. **MemnotWrD0**, **MemnotRfD1** and **MemAD2-32** are all high at the beginning of the scan. Starting with **MemnotWrD0**, each of these lines goes low successively at intervals of two **ClockIn** periods and stays low until the end of the scan. If one of these lines is connected to **MemConfig** the preset internal configuration mode associated with that line will be used as the EMI configuration. The default configuration is that defined in the table for **MemAD31**; connecting **MemConfig** to **VCC** will also produce this default configuration. Note that only 13 of the possible configurations are valid.

Table 7.6: IMS T414 internal configuration coding

Pin	Duration of each Tstate periods Tm						Strobe coefficient				Write cycle	Refresh interval	Cycle time	Extra cycles
	T1	T2	T3	T4	T5	T6	s1	s2	s3	s4	type	ClockIn cycles	Proc cycles	e
MemnotWrD0	1	1	1	1	1	1	30	1	3	5	late	72	3	2
MemnotRfD1	1	2	1	1	1	2	30	1	2	7	late	72	4	3
MemAD2	1	2	1	1	2	3	30	1	2	7	late	72	5	4
MemAD3	2	3	1	1	2	3	30	1	3	8	late	72	6	5
MemAD4	1	1	1	1	1	1	3	1	2	3	early	72	3	2
MemAD5	1	1	2	1	2	1	5	1	2	3	early	72	4	3
MemAD6	2	1	2	1	3	1	6	1	2	3	early	72	5	4
MemAD7	2	2	2	1	3	2	7	1	3	4	early	72	6	5
MemAD8	1	1	1	1	1	1	30	1	2	3	early	—	3	2
MemAD9	1	1	2	1	2	1	30	2	5	9	early	—	4	3
MemAD10	2	2	2	2	4	2	30	2	3	8	late	72	7	6
MemAD11	3	3	3	3	3	3	30	2	4	13	late	72	9	8
MemAD31	4	4	4	4	4	4	31	30	30	18	late	72	12	11

Table 7.7: IMS T414 internal configuration description

Pin	Configuration
MemnotWrD0	Dynamic RAM in 3 processor cycles
MemnotRfD1	Dynamic RAM in 4 processor cycles
MemAD2	Dynamic RAM in 5 processor cycles
MemAD3	Dynamic RAM in 6 cycles
MemAD4	Multiplexed address dynamic RAM in 3 processor cycles
MemAD5	Multiplexed address dynamic RAM in 4 processor cycles
MemAD6	Multiplexed address dynamic RAM in 5 processor cycles
MemAD7	Multiplexed address dynamic RAM in 6 processor cycles
MemAD8	Fast static RAM in 3 processor cycles
MemAD9	Static RAM in 4 cycles with wait generator
MemAD10	General purpose configuration in 7 processor cycles
MemAD11	General purpose configuration in 9 processor cycles
MemAD31	General purpose configuration in 12 processor cycles

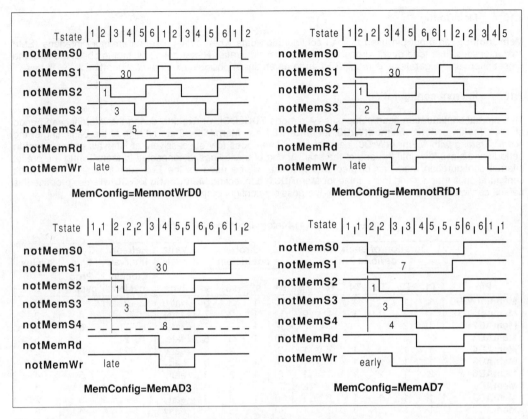

Figure 7.9: IMS T414 internal configuration

7.10.2 External configuration

If **MemConfig** is held low until **MemnotWrD0** goes low the internal configuration is ignored and an external configuration will be loaded instead. An external configuration scan always follows an internal one, but if an internal configuration occurs any external configuration is ignored.

The external configuration scan comprises 36 successive external read cycles, using the default EMI configuration preset by **MemAD31**. However, instead of data being read on the data bus as for a normal read cycle, only a single bit of data is read on **MemConfig** at each cycle. Addresses put out on the bus for each read cycle are shown in table 7.8, and are designed to address ROM at the top of the memory map. The table shows the data to be held in ROM; data required at the **MemConfig** pin is the inverse of this.

MemConfig is typically connected via an inverter to **MemnotWrD0**. Data bit zero of the least significant byte of each ROM word then provides the configuration data stream. By switching **MemConfig** between various data bus lines up to 32 configurations can be stored in ROM, one per bit of the data bus. **MemConfig** can be permanently connected to a data line or to **GND**. Connecting **MemConfig** to **GND** gives all **Tstates** configured to four periods; **notMemS1** pulse of maximum duration; **notMemS2-4** delayed by maximum; refresh interval 72 periods of **ClockIn**; refresh enabled; late write.

The external memory configuration table 7.8 shows the contribution of each memory address to the 13 configuration fields. The lowest 12 words (#7FFFFF6C to #7FFFFF98, fields 1 to 6) define the number of extra periods **Tm** to be added to each **Tstate**. If field 2 is 3 then three extra periods will be added to **T2** to extend it to the maximum of four periods.

7 External memory interface 143

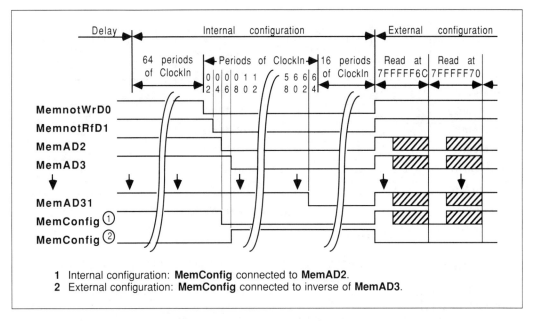

Figure 7.10: IMS T414 internal configuration scan

The next five addresses (field 7) define the duration of **notMemS1** and the following fifteen (fields 8 to 10) define the delays before strobes **notMemS2-4** become active. The five bits allocated to each strobe allow durations of from 0 to 31 periods **Tm**, as described in strobes page 134.

Addresses #7FFFFFEC to #7FFFFFF4 (fields 11 and 12) define the refresh interval and whether refresh is to be used, whilst the final address (field 13) supplies a high bit to **MemConfig** if a late write cycle is required.

The columns to the right of the coding table show the values of each configuration bit for the four sample external configuration diagrams. Note the inclusion of period **E** at the end of **T6** in some diagrams. This is inserted to bring the start of the next **Tstate T1** to coincide with a rising edge of **ProcClockOut** (page 132).

Wait states **W** have been added to show the effect of them on strobe timing; they are not part of a configuration. In each case which includes wait states, two wait periods are defined. This shows that if a wait state would cause the start of **T5** to coincide with a falling edge of **ProcClockOut**, another period **Tm** is generated by the EMI to force it to coincide with a rising edge of **ProcClockOut**. This coincidence is only necessary if wait states are added, otherwise coincidence with a falling edge is permitted.

144　　　　　　　　　　　　　　　　　　　　　4　IMS T414 engineering data

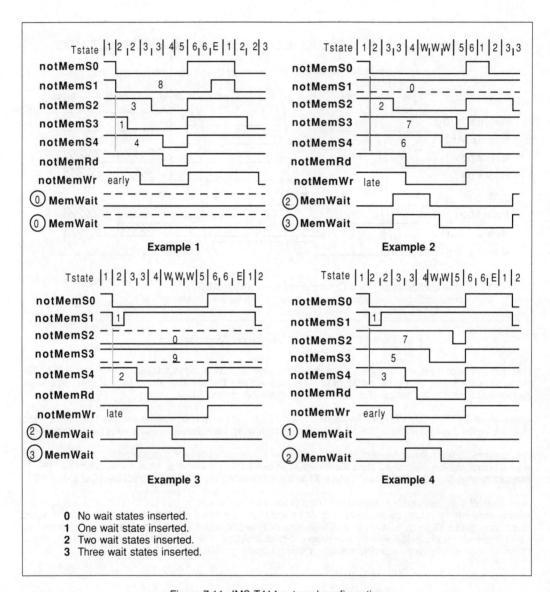

Figure 7.11: IMS T414 external configuration

7 External memory interface

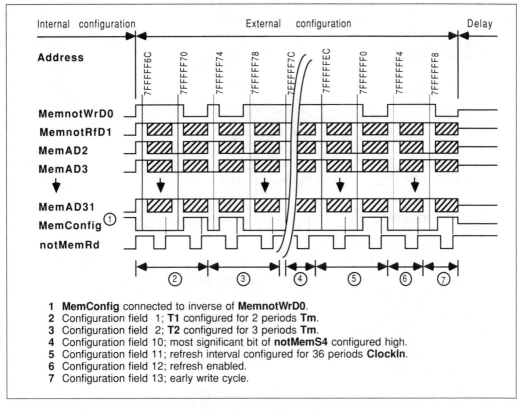

Figure 7.12: IMS T414 external configuration scan

Table 7.8: IMS T414 external configuration coding

Scan cycle	MemAD address	Field	Function	Example diagram 1	2	3	4
1	7FFFFF6C	1	T1 least significant bit	0	0	0	0
2	7FFFFF70	1	T1 most significant bit	0	0	0	0
3	7FFFFF74	2	T2 least significant bit	1	0	0	1
4	7FFFFF78	2	T2 most significant bit	0	0	0	0
5	7FFFFF7C	3	T3 least significant bit	1	1	1	1
6	7FFFFF80	3	T3 most significant bit	0	0	0	0
7	7FFFFF84	4	T4 least significant bit	0	0	0	0
8	7FFFFF88	4	T4 most significant bit	0	0	0	0
9	7FFFFF8C	5	T5 least significant bit	0	0	0	0
10	7FFFFF90	5	T5 most significant bit	0	0	0	0
11	7FFFFF94	6	T6 least significant bit	1	0	1	1
12	7FFFFF98	6	T6 most significant bit	0	0	0	0
13	7FFFFF9C	7	notMemS1 least significant bit	0	0	1	1
14	7FFFFFA0	7		0	0	0	0
15	7FFFFFA4	7	⇓ ⇓	0	0	0	0
16	7FFFFFA8	7		1	0	0	0
17	7FFFFFAC	7	notMemS1 most significant bit	0	0	0	0
18	7FFFFFB0	8	notMemS2 least significant bit	1	0	0	1
19	7FFFFFB4	8		1	1	0	1
20	7FFFFFB8	8	⇓ ⇓	0	0	0	1
21	7FFFFFBC	8		0	0	0	0
22	7FFFFFC0	8	notMemS2 most significant bit	0	0	0	0
23	7FFFFFC4	9	notMemS3 least significant bit	1	1	1	1
24	7FFFFFC8	9		0	1	0	0
25	7FFFFFCC	9	⇓ ⇓	0	1	0	1
26	7FFFFFD0	9		0	0	1	0
27	7FFFFFD4	9	notMemS3 most significant bit	0	0	0	0
28	7FFFFFD8	10	notMemS4 least significant bit	0	0	0	1
29	7FFFFFDC	10		0	1	1	1
30	7FFFFFE0	10	⇓ ⇓	1	1	0	0
31	7FFFFFE4	10		0	0	0	0
32	7FFFFFE8	10	notMemS4 most significant bit	0	0	0	0
33	7FFFFFEC	11	Refresh Interval least significant bit	-	-	-	-
34	7FFFFFF0	11	Refresh Interval most significant bit	-	-	-	-
35	7FFFFFF4	12	Refresh Enable	-	-	-	-
36	7FFFFFF8	13	Late Write	0	1	1	0

Table 7.9: IMS T414 memory refresh configuration coding

Refresh interval	Interval in μs	Field 11 encoding	Complete cycle (mS)
18	3.6	00	0.922
36	7.2	01	1.843
54	10.8	10	2.765
72	14.4	11	3.686

Refresh intervals are in periods of **ClockIn** and **ClockIn** frequency is 5MHz:

$$\text{Interval} = 18 * 200 = 3600\text{ns}$$

Refresh interval is between successive incremental refresh addresses.
Complete cycles are shown for 256 row DRAMS.

7 External memory interface

Table 7.10: Memory configuration

SYMBOL	PARAMETER	MIN	NOM	MAX	UNITS	NOTE
TMCVRdH	Memory configuration data setup	20			ns	
TRdHMCX	Memory configuration data hold	0			ns	
TS0LRdH	notMemS0 to configuration data read	a		a+6	ns	1

Notes

1 **a** is 16 periods **Tm**.

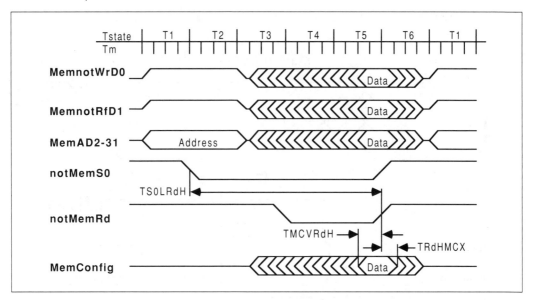

Figure 7.13: IMS T414 external configuration read cycle timing

7.11   notMemRf

The IMS T414 can be operated with memory refresh enabled or disabled. The selection is made during memory configuration, when the refresh interval is also determined. Refresh cycles do not interrupt internal memory accesses, although the internal addresses cannot be reflected on the external bus during refresh.

When refresh is disabled no refresh cycles occur. During the post-**Reset** period eight dummy refresh cycles will occur with the appropriate timing but with no bus or strobe activity.

A refresh cycle uses the same basic external memory timing as a normal external memory cycle, except that it starts two periods **Tm** before the start of **T1**. If a refresh cycle is due during an external memory access, it will be delayed until the end of that external cycle. Two extra periods **Tm** (periods **R** in the diagram) will then be inserted between the end of **T6** of the external memory cycle and the start of **T1** of the refresh cycle itself. The refresh address and various external strobes become active approximately one period **Tm** before **T1**. Bus signals are active until the end of **T2**, whilst **notMemRf** remains active until the end of **T6**.

For a refresh cycle, **MemnotRfD1** goes low before **notMemRf** goes low and **MemnotWrD0** goes high with the same timing as **MemnotRfD1**. All the address lines share the same timing, but only **MemAD2-11** give the refresh address. **MemAD12-30** stay high during the address period, whilst **MemAD31** remains low. Refresh cycles generate strobes **notMemS0-4** with timing as for a normal external cycle, but **notMemRd** and **notMemWrB0-3** remain high. **MemWait** operates normally during refresh cycles.

Table 7.11: Memory refresh

SYMBOL	PARAMETER	MIN	NOM	MAX	UNITS	NOTE
TRfLRfH	Refresh pulse width low	a		a+6	ns	1
TRaVS0L	Refresh address setup before notMemS0		b		ns	2
TRfLS0L	Refresh indicator setup before notMemS0		b		ns	2

Notes

1 **a** is total **Tmx**+(2 periods **Tm**).

2 **b** is total **T1**+(2 periods **Tm**) where **T1** can be from one to four periods **Tm** in length.

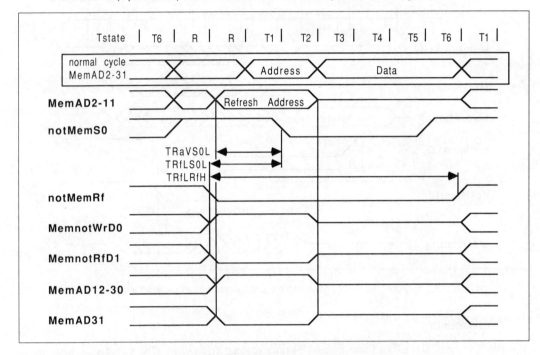

Figure 7.14: IMS T414 refresh cycle timing

7.12 MemWait

Taking **MemWait** high with the timing shown will extend the duration of **T4**. **MemWait** is sampled near to, but independent of, the falling edge of **ProcClockOut**, and should not change state in this region. By convention, **notMemS4** is used to synchronize wait state insertion. If this or another strobe is used, its delay should be such as to take the strobe low an even number of periods **Tm** after the start of **T1**, to coincide with a rising edge of **ProcClockOut**.

MemWait may be kept high indefinitely, although if dynamic memory refresh is used it should not be kept high long enough to interfere with refresh timing. **MemWait** operates normally during all cycles, including refresh and configuration cycles.

If the start of **T5** would coincide with a falling edge of **ProcClockOut** an extra wait period **Tm** (**EW**) is generated by the EMI to force coincidence with a rising edge. Rising edge coincidence is only forced if wait states are added, otherwise coincidence with a falling edge is permitted.

7 External memory interface

Table 7.12: Memory wait

SYMBOL	PARAMETER	MIN	NOM	MAX	UNITS	NOTE
TPCHWtH	Wait setup	-(a)+3			ns	1,4
TPCHWtL	Wait hold	b+3			ns	2,3,4
TWtLWtH	Delay before re-assertion of Wait	2			Tm	

Notes

1. **a** is 0.5 periods **Tm**.

2. **b** is 1.5 periods **Tm**.

3. If wait period exceeds refresh interval, refresh cycles will be lost.

4. Wait timing is independent of falling edge of **ProcClockOut**.

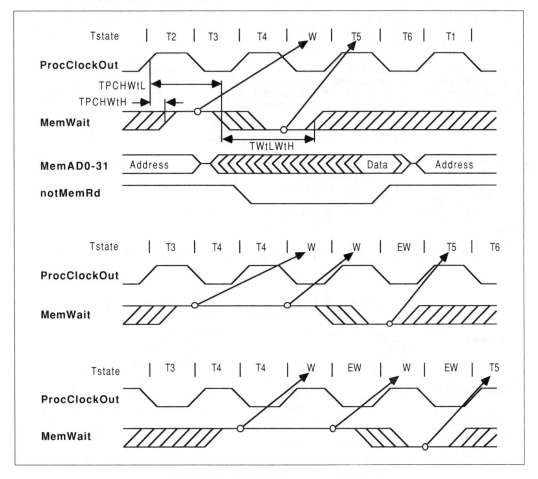

Figure 7.15: IMS T414 memory wait timing

7.13 MemReq, MemGranted

Direct memory access (DMA) can be requested at any time by taking the asynchronous **MemReq** input high. The transputer samples **MemReq** during the final period **Tm** of **T6** of both refresh and external memory cycles. To guarantee taking over the bus immediately following either, **MemReq** must be set up at least two periods **Tm** before the end of **T6**. In the absence of an external memory cycle, **MemReq** is sampled during every low period of **ProcClockOut**. The address bus is tristated two periods **Tm** after the **ProcClockOut** rising edge which follows the sample. **MemGranted** is asserted one period **Tm** after that.

Removal of **MemReq** is sampled during each low period of **ProcClockOut** and **MemGranted** is removed synchronously with the next falling edge of **ProcClockOut**. If accurate timing of DMA is required, **MemReq** should be set low coincident with a falling edge of **ProcClockOut**. Further external bus activity, either refresh, external cycles or reflection of internal cycles, will commence at the next rising edge of **ProcClockOut**.

Strobes are left in their inactive states during DMA. DMA cannot interrupt a refresh or external memory cycle, and outstanding refresh cycles will occur before the bus is released to DMA. DMA does not interfere with internal memory cycles in any way, although a program running in internal memory would have to wait for the end of DMA before accessing external memory. DMA cannot access internal memory. If DMA extends longer than one refresh interval (Memory Refresh Configuration Coding table, page 142), the DMA user becomes responsible for refresh. DMA may also inhibit an internally running program from accessing external memory.

DMA allows a bootstrap program to be loaded into external RAM ready for execution after reset. If **MemReq** is held high throughout reset, **MemGranted** will be asserted before the bootstrap sequence begins. **MemReq** must be high at least one period **TDCLDCL** of **ClockIn** before **Reset**. The circuit should be designed to ensure correct operation if **Reset** could interrupt a normal DMA cycle.

Table 7.13: Memory request

SYMBOL	PARAMETER	MIN	NOM	MAX	UNITS	NOTE
TMRHMGH	Memory request response time	4		6	Tm	1
TMRLMGL	Memory request end response time	2		4	Tm	
TADZMGH	Bus tristate before memory granted		1		Tm	
TMGLADV	Bus active after end of memory granted		1		Tm	

Notes

1 These values assume no external memory cycle is in progress. If an external cycle is active, maximum time could be (1 EMI cycle **Tmx**)+(1 refresh cycle **TRfLRfH**)+(6 periods **Tm**).

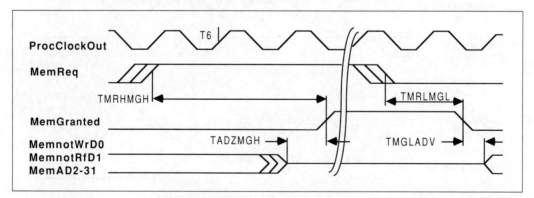

Figure 7.16: IMS T414 memory request timing

7 External memory interface

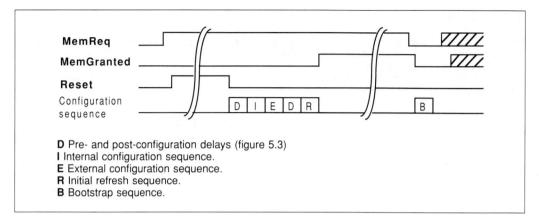

Figure 7.17: IMS T414 DMA sequence at reset

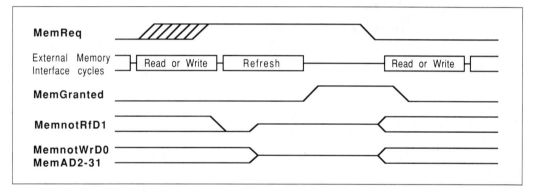

Figure 7.18: IMS T414 operation of MemReq, MemGranted with external, refresh memory cycles

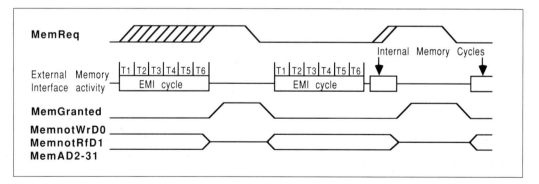

Figure 7.19: IMS T414 operation of MemReq, MemGranted with external, internal memory cycles

8   Events

EventReq and **EventAck** provide an asynchronous handshake interface between an external event and an internal process. When an external event takes **EventReq** high the external event channel (additional to the external link channels) is made ready to communicate with a process. When both the event channel and the process are ready the processor takes **EventAck** high and the process, if waiting, is scheduled. **EventAck** is removed after **EventReq** goes low.

Only one process may use the event channel at any given time. If no process requires an event to occur **EventAck** will never be taken high. Although **EventReq** triggers the channel on a transition from low to high, it must not be removed before **EventAck** is high. **EventReq** should be low during **Reset**; if not it will be ignored until it has gone low and returned high. **EventAck** is taken low when **Reset** occurs.

If the process is a high priority one and no other high priority process is running, the latency is as described on page 115. Setting a high priority task to wait for an event input is a way of interrupting a transputer program.

Table 8.1: Event

SYMBOL	PARAMETER	MIN	NOM	MAX	UNITS	NOTE
TVHKH	Event request response	0			ns	
TKHVL	Event request hold	0			ns	
TVLKL	Delay before removal of event acknowledge	0		a	ns	1
TKLVH	Delay before re-assertion of event request	0			ns	

Notes

1   **a** is **TPCLPCL** (2 periods **Tm**).

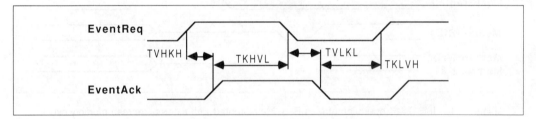

Figure 8.1: IMS T414 event timing

9 Links

Four identical INMOS bi-directional serial links provide synchronized communication between processors and with the outside world. Each link comprises an input channel and output channel. A link between two transputers is implemented by connecting a link interface on one transputer to a link interface on the other transputer Every byte of data sent on a link is acknowledged on the input of the same link, thus each signal line carries both data and control information.

The quiescent state of a link output is low. Each data byte is transmitted as a high start bit followed by a one bit followed by eight data bits followed by a low stop bit. The least significant bit of data is transmitted first. After transmitting a data byte the sender waits for the acknowledge, which consists of a high start bit followed by a zero bit. The acknowledge signifies both that a process was able to receive the acknowledged data byte and that the receiving link is able to receive another byte. The sending link reschedules the sending process only after the acknowledge for the final byte of the message has been received.

The IMS T414 links support the standard INMOS communication speed of 10 Mbits per second. In addition they can be used at 5 or 20 Mbits per second. Links are not synchronised with **ClockIn** or **ProcClockOut** and are insensitive to their phases. Thus links from independently clocked systems may communicate, providing only that the clocks are nominally identical and within specification.

Links are TTL compatible and intended to be used in electrically quiet environments, between devices on a single printed circuit board or between two boards via a backplane. Direct connection may be made between devices separated by a distance of less than 300 millimetres. For longer distances a matched 100 Ohm transmission line should be used with series matching resistors **RM**. When this is done the line delay should be less than 0.4 bit time to ensure that the reflection returns before the next data bit is sent.

Buffers may be used for very long transmissions. If so, their overall propagation delay should be stable within the skew tolerance of the link, although the absolute value of the delay is immaterial.

Link speeds can be set by **LinkSpecial**, **Link0Special** and **Link123Special**. The link 0 speed can be set independently. Table 9.1 shows uni-directional and bi-directional data rates in Kbytes/second for each link speed; **LinknSpecial** is to be read as **Link0Special** when selecting link 0 speed and as **Link123Special** for the others. Data rates are quoted for a transputer using internal memory, and will be affected by a factor depending on the number of external memory accesses and the length of the external memory cycle.

Table 9.1: Speed Settings for Transputer Links

Link Special	Linkn Special	Mbits/sec	Kbytes/sec	
			Uni	Bi
0	0	10	400	800
0	1	5	200	400
1	0	10	400	800
1	1	20	800	1600

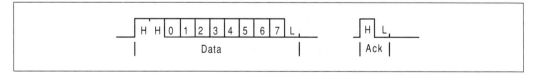

Figure 9.1: IMS T414 link data and acknowledge packets

Table 9.2: Link

SYMBOL	PARAMETER		MIN	NOM	MAX	UNITS	NOTE
TJQr	LinkOut rise time				20	ns	
TJQf	LinkOut fall time				10	ns	
TJDr	LinkIn rise time				20	ns	
TJDf	LinkIn fall time				20	ns	
TJQJD	Buffered edge delay		0			ns	
TJBskew	Variation in TJQJD	20 Mbits/s			3	ns	1
		10 Mbits/s			10	ns	1
		5 Mbits/s			30	ns	1
CLIZ	LinkIn capacitance	@ f=1MHz			7	pF	
CLL	LinkOut load capacitance				50	pF	
RM	Series resistor for 100Ω transmission line			56		ohms	

Notes

1 This is the variation in the total delay through buffers, transmission lines, differential receivers etc., caused by such things as short term variation in supply voltages and differences in delays for rising and falling edges.

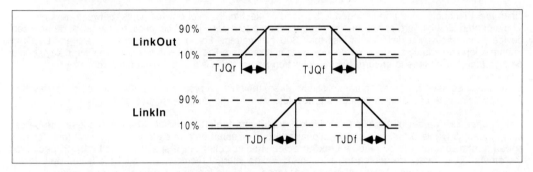

Figure 9.2: IMS T414 link timing

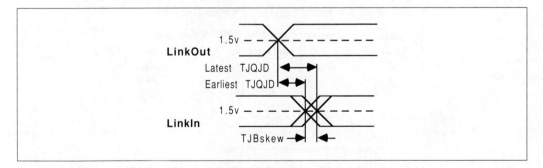

Figure 9.3: IMS T414 buffered link timing

9 Links

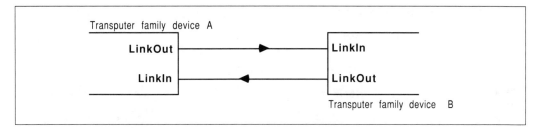

Figure 9.4: Links directly connected

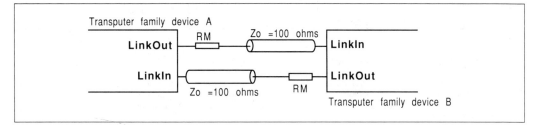

Figure 9.5: Links connected by transmission line

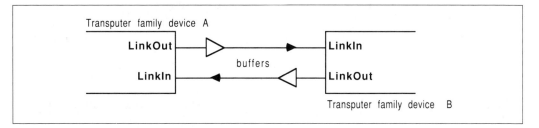

Figure 9.6: Links connected by buffers

10 Electrical specifications

10.1 DC electrical characteristics

Table 10.1: Absolute maximum ratings

SYMBOL	PARAMETER	MIN	MAX	UNITS	NOTE
VCC	DC supply voltage	0	7.0	V	1,2,3
VI, VO	Voltage on input and output pins	-0.5	VCC+0.5	V	1,2,3
II	Input current		±25	mA	4
OSCT	Output short circuit time (one pin)		1	s	2
TS	Storage temperature	-65	150	°C	2
TA	Ambient temperature under bias	-55	125	°C	2
PDmax	Maximum allowable dissipation		2	W	

Notes

1. All voltages are with respect to **GND**.

2. This is a stress rating only and functional operation of the device at these or any other conditions beyond those indicated in the operating sections of this specification is not implied. Stresses greater than those listed may cause permanent damage to the device. Exposure to absolute maximum rating conditions for extended periods may affect reliability.

3. This device contains circuitry to protect the inputs against damage caused by high static voltages or electrical fields. However, it is advised that normal precautions be taken to avoid application of any voltage higher than the absolute maximum rated voltages to this high impedance circuit. Unused inputs should be tied to an appropriate logic level such as **VCC** or **GND**.

4. The input current applies to any input or output pin and applies when the voltage on the pin is between **GND** and **VCC**.

Table 10.2: Operating conditions

SYMBOL	PARAMETER	MIN	MAX	UNITS	NOTE
VCC	DC supply voltage	4.75	5.25	V	1
VI, VO	Input or output voltage	0	VCC	V	1,2
CL	Load capacitance on any pin		50	pF	
TA	Operating temperature range	0	70	°C	3

Notes

1. All voltages are with respect to **GND**.

2. Excursions beyond the supplies are permitted but not recommended; see DC characteristics.

3. Air flow rate 400 linear ft/min transverse air flow.

10 Electrical specifications

Table 10.3: DC characteristics

SYMBOL	PARAMETER		MIN	MAX	UNITS	NOTE
VIH	High level input voltage		2.0	VCC+0.5	V	1,2
VIL	Low level input voltage		-0.5	0.8	V	1,2
II	Input current	@ GND<VI<VCC		±10	µA	1,2,6
				±50	µA	1,2,7
VOH	Output high voltage	@ IOH=2mA	VCC-1		V	1,2
VOL	Output low voltage	@ IOL=4mA		0.4	V	1,2
IOS	Output short circuit current	@ GND<VO<VCC		50	mA	1,2,4
				75	mA	1,2,5
IOZ	Tristate output current	@ GND<VO<VCC		±10	µA	1,2
PD	Power dissipation			900	mW	2,3
CIN	Input capacitance	@ f=1MHz		7	pF	
COZ	Output capacitance	@ f=1MHz		10	pF	

Notes

1. All voltages are with respect to **GND**.

2. Parameters measured at 4.75V<**VCC**<5.25V and 0°C<**TA**<70°C. Input clock frequency = 5MHz.

3. Power dissipation varies with output loading and program execution.

4. Current sourced from non-link outputs.

5. Current sourced from link outputs.

6. For inputs other than those in Note 7.

7. For **MemReq**, **MemWait**, **MemConfig**, **Analyse**, **Reset**, **ClockIn**, **EventReq**, **LinkIn0-3**, **LinkSpecial**, **Link0Special**, **Link123Special**, **BootFromRom**, **HoldToGND**.

10.2 Equivalent circuits

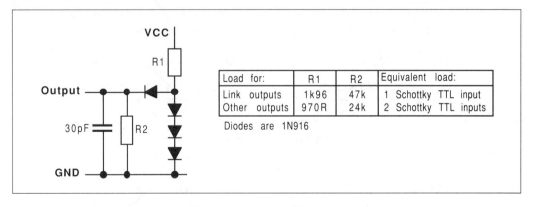

Figure 10.1: Load circuit for AC measurements

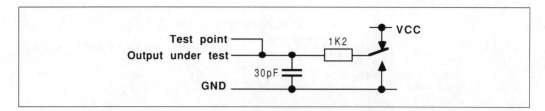

Figure 10.2: Tristate load circuit for AC measurements

10.3 AC timing characteristics

Table 10.4: Input, output edges

SYMBOL	PARAMETER	MIN	MAX	UNITS	NOTE
TDr	Input rising edges	2	20	ns	1,2
TDf	Input falling edges	2	20	ns	1,2
TQr	Output rising edges		25	ns	1
TQf	Output falling edges		15	ns	1
TS0LaHZ	Address high to tristate	a	a+6	ns	3
TS0LaLZ	Address low to tristate	a	a+6	ns	3

Notes

1 Non-link pins; see section on links.

2 All inputs except **ClockIn**; see section on **ClockIn**.

3 **a** is **T2** where **T2** can be from one to four periods **Tm** in length.
Address lines include **MemnotWrD0**, **MemnotRfD1**, **MemAD2-31**.

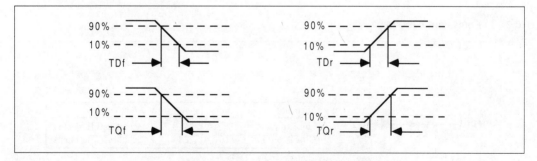

Figure 10.3: IMS T414 input and output edge timing

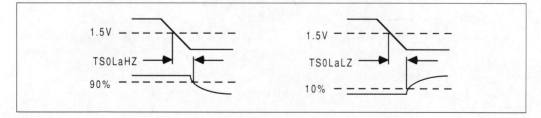

Figure 10.4: IMS T414 tristate timing relative to notMemS0

10 Electrical specifications

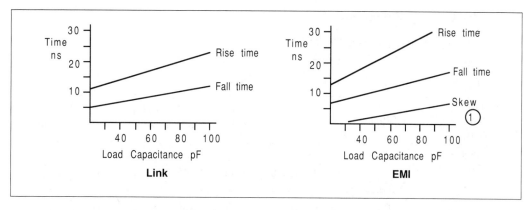

Figure 10.5: Typical rise/fall times

Notes

1. Skew is measured between **notMemS0** with a standard load (2 Schottky TTL inputs and 30pF) and **notMemS0** with a load of 2 Schottky TTL inputs and varying capacitance.

10.4 Power rating

Internal power dissipation P_{INT} of transputer and peripheral chips depends on **VCC**, as shown in figure 10.6. P_{INT} is substantially independent of temperature.

Total power dissipation P_D of the chip is

$$P_D = P_{INT} + P_{IO}$$

where P_{IO} is the power dissipation in the input and output pins; this is application dependent.

Internal working temperature T_J of the chip is

$$T_J = T_A + \theta J_A * P_D$$

where T_A is the external ambient temperature in °C and θJ_A is the junction-to-ambient thermal resistance in °C/W. θJ_A for each package is given in the Packaging Specifications section.

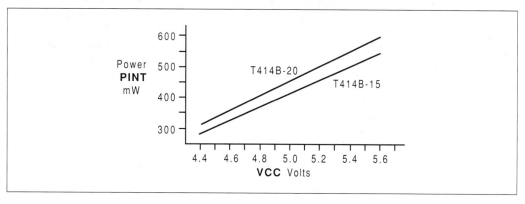

Figure 10.6: IMS T414 internal power dissipation vs VCC

11 Package specifications

11.1 84 pin grid array package

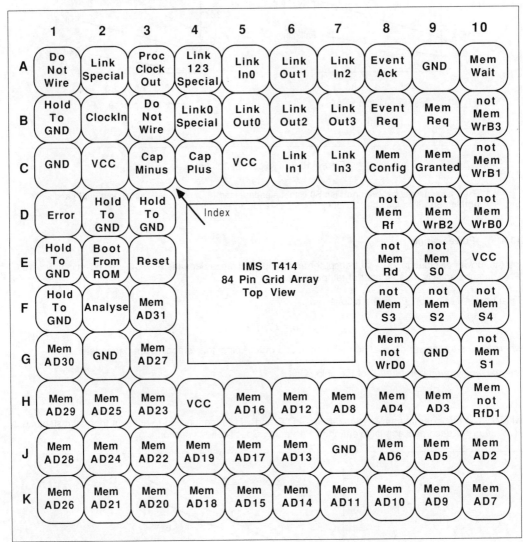

Figure 11.1: IMS T414 84 pin grid array package pinout

11 Package specifications

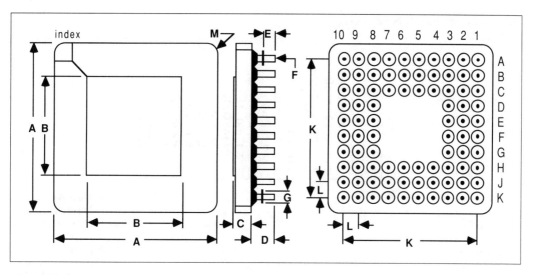

Figure 11.2: 84 pin grid array package dimensions

Table 11.1: 84 pin grid array package dimensions

DIM	Millimetres		Inches		Notes
	NOM	TOL	NOM	TOL	
A	26.924	±0.254	1.060	±0.010	
B	17.019	±0.127	0.670	±0.005	
C	2.456	±0.278	0.097	±0.011	
D	4.572	±0.127	0.180	±0.005	
E	3.302	±0.127	0.130	±0.005	
F	0.457	±0.025	0.018	±0.001	Pin diameter
G	1.143	±0.127	0.045	±0.005	Flange diameter
K	22.860	±0.127	0.900	±0.005	
L	2.540	±0.127	0.100	±0.005	
M	0.508		0.020		Chamfer

Package weight is approximately 7.2 grams

Table 11.2: 84 pin grid array package junction to ambient thermal resistance

SYMBOL	PARAMETER	MIN	NOM	MAX	UNITS	NOTE
θ_{JA}	At 400 linear ft/min transverse air flow			35	°C/W	

11.1.1 84 pin PLCC J-bend package

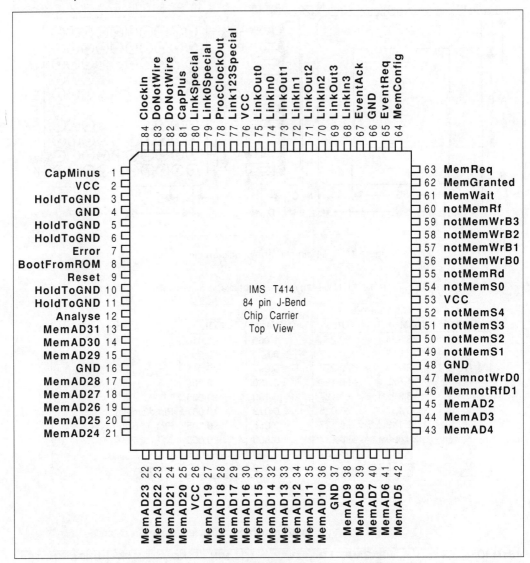

Figure 11.3: IMS T414 84 pin PLCC J-bend package pinout

11 Package specifications

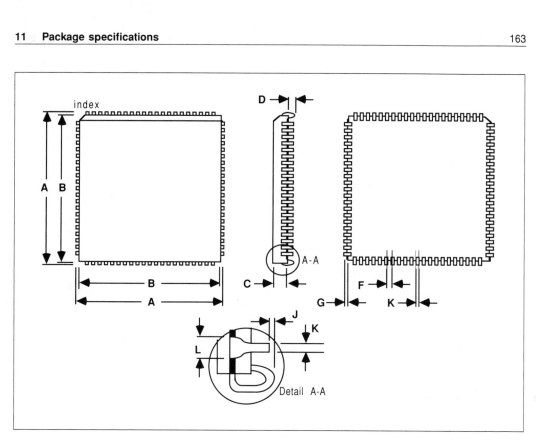

Figure 11.4: 84 pin PLCC J-bend package dimensions

Table 11.3: 84 pin PLCC J-bend package dimensions

DIM	Millimetres		Inches		Notes
	NOM	TOL	NOM	TOL	
A	30.226	±0.127	1.190	±0.005	
B	29.312	±0.127	1.154	±0.005	
C	3.810	±0.127	0.150	±0.005	
D	0.508	±0.127	0.020	±0.005	
F	1.270	±0.127	0.050	±0.005	
G	0.457	±0.127	0.018	±0.005	
J	0.000	±0.051	0.000	±0.002	
K	0.457	±0.127	0.018	±0.005	
L	0.762	±0.127	0.030	±0.005	

Package weight is approximately 7.0 grams

Table 11.4: 84 pin PLCC J-bend package junction to ambient thermal resistance

SYMBOL	PARAMETER	MIN	NOM	MAX	UNITS	NOTE
θJA	At 400 linear ft/min transverse air flow		35		°C/W	

Chapter 5

IMS T212 engineering data

1 Introduction

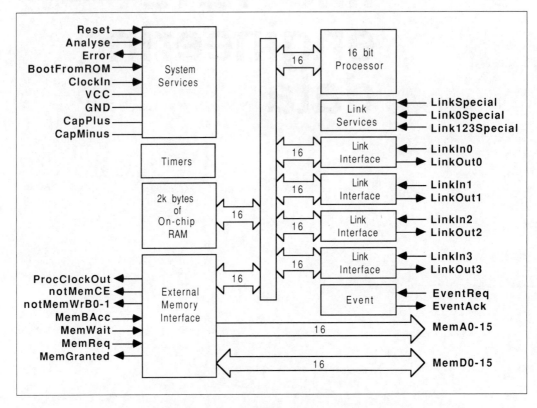

Figure 1.1: IMS T212 block diagram

1 Introduction

The IMS T212 transputer is a 16 bit CMOS microcomputer with 2 Kbytes on-chip RAM for high speed processing, an external memory interface and four standard INMOS communication links. The instruction set achieves efficient implementation of high level languages and provides direct support for the occam model of concurrency when using either a single transputer or a network. Procedure calls, process switching and typical interrupt latency are sub-microsecond. A device running at 20 MHz achieves an instruction throughput of 10 MIPS.

The IMS T212 can directly access a linear address space of 64 Kbytes. The 16 bit wide non-multiplexed external memory interface provides a data rate of up to 2 bytes every 100 nanoseconds (20 Mbytes/sec) for a 20 MHz device. System Services include processor reset and bootstrap control, together with facilities for error analysis.

The INMOS communication links allow networks of transputers to be constructed by direct point to point connections with no external logic. The links support the standard operating speed of 10 Mbits per second, but also operate at 5 or 20 Mbits per second.

The IMS T212 is designed to implement the occam language, detailed in the occam Reference Manual, but also efficiently supports other languages such as C and Pascal. Access to specific features of the IMS T212 is described in the relevant system development manual. Access to the transputer at machine level is seldom required, but if necessary refer to The Transputer Instruction Set - A Compiler Writers' Guide.

This data sheet supplies hardware implementation and characterisation details for the IMS T212. It is intended to be read in conjunction with the Transputer Architecture chapter, which details the architecture of the transputer and gives an overview of occam. For convenience of description, the IMS T212 operation is split into the basic blocks shown in figure 1.1.

2 Pin designations

Table 2.1: IMS T212 system services

Pin	In/Out	Function
VCC, GND		Power supply and return
CapPlus, CapMinus		External capacitor for internal clock power supply
ClockIn	in	Input clock
Reset	in	System reset
Error	out	Error indicator
Analyse	in	Error analysis
BootFromRom	in	Bootstraps from external ROM or from link
HoldToGND		Must be connected to **GND**

Table 2.2: IMS T212 external memory interface

Pin	In/Out	Function
ProcClockOut	out	Processor clock
MemA0-15	out	Sixteen address lines
MemD0-15	in/out	Sixteen data lines
notMemWrB0-1	out	Two byte-addressing write strobes
notMemCE	out	Chip enable
MemBAcc	in	Byte access mode selector
MemWait	in	Memory cycle extender
MemReq	in	Direct memory access request
MemGranted	out	Direct memory access granted

Table 2.3: IMS T212 event

Pin	In/Out	Function
EventReq	in	Event request
EventAck	out	Event request acknowledge

Table 2.4: IMS T212 link

Pin	In/Out	Function
LinkIn0-3	in	Four serial data input channels
LinkOut0-3	out	Four serial data output channels
LinkSpecial	in	Select non-standard speed as 5 or 20 Mbits/sec
Link0Special	in	Select special speed for Link 0
Link123Special	in	Select special speed for Links 1,2,3

Signal names are prefixed by **not** if they are active low, otherwise they are active high.
Pinout details for various packages are given on page 208.

3 Processor

The 16 bit processor contains instruction processing logic, instruction and work pointers, and an operand register. It directly accesses the high speed 2 Kbyte on-chip memory, which can store data or program. Where larger amounts of memory or programs in ROM are required, the processor has access to 64 Kbytes of memory via the External Memory Interface (EMI).

3.1 Registers

The design of the transputer processor exploits the availability of fast on-chip memory by having only a small number of registers; six registers are used in the execution of a sequential process. The small number of registers, together with the simplicity of the instruction set, enables the processor to have relatively simple (and fast) data-paths and control logic. The six registers are:

> The workspace pointer which points to an area of store where local variables are kept.
>
> The instruction pointer which points to the next instruction to be executed.
>
> The operand register which is used in the formation of instruction operands.
>
> The *A*, *B* and *C* registers which form an evaluation stack.

A, *B* and *C* are sources and destinations for most arithmetic and logical operations. Loading a value into the stack pushes *B* into *C*, and *A* into *B*, before loading *A*. Storing a value from *A*, pops *B* into *A* and *C* into *B*.

Expressions are evaluated on the evaluation stack, and instructions refer to the stack implicitly. For example, the *add* instruction adds the top two values in the stack and places the result on the top of the stack. The use of a stack removes the need for instructions to respecify the location of their operands. Statistics gathered from a large number of programs show that three registers provide an effective balance between code compactness and implementation complexity.

No hardware mechanism is provided to detect that more than three values have been loaded onto the stack. It is easy for the compiler to ensure that this never happens.

Any location in memory can be accessed relative to the workpointer register, enabling the workspace to be of any size.

Further register details are given in The Transputer Instruction Set - A Compiler Writers' Guide.

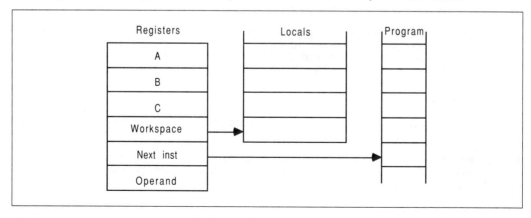

Figure 3.1: Registers

3.2 Instructions

The instruction set has been designed for simple and efficient compilation of high-level languages. All instructions have the same format, designed to give a compact representation of the operations occurring most frequently in programs.

Each instruction consists of a single byte divided into two 4-bit parts. The four most significant bits of the byte are a function code and the four least significant bits are a data value.

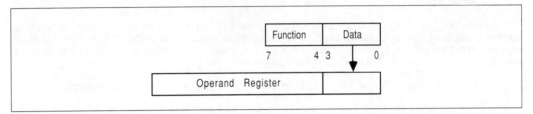

Figure 3.2: Instruction format

3.2.1 Direct functions

The representation provides for sixteen functions, each with a data value ranging from 0 to 15. Thirteen of these, shown in table 3.1, are used to encode the most important functions.

Table 3.1: Direct functions

load constant	add constant	
load local	store local	load local pointer
load non-local	store non-local	
jump	conditional jump	call

The most common operations in a program are the loading of small literal values and the loading and storing of one of a small number of variables. The *load constant* instruction enables values between 0 and 15 to be loaded with a single byte instruction. The *load local* and *store local* instructions access locations in memory relative to the workspace pointer. The first 16 locations can be accessed using a single byte instruction.

The *load non-local* and *store non-local* instructions behave similarly, except that they access locations in memory relative to the *A* register. Compact sequences of these instructions allow efficient access to data structures, and provide for simple implementations of the static links or displays used in the implementation of high level programming languages such as occam, C or Pascal.

3.2.2 Prefix functions

Two more function codes allow the operand of any instruction to be extended in length; *prefix* and *negative prefix*.

All instructions are executed by loading the four data bits into the least significant four bits of the operand register, which is then used as the instruction's operand. All instructions except the prefix instructions end by clearing the operand register, ready for the next instruction.

The *prefix* instruction loads its four data bits into the operand register and then shifts the operand register up four places. The *negative prefix* instruction is similar, except that it complements the operand register before shifting it up. Consequently operands can be extended to any length up to the length of the operand register by a sequence of prefix instructions. In particular, operands in the range -256 to 255 can be represented using one prefix instruction.

3 Processor

The use of prefix instructions has certain beneficial consequences. Firstly, they are decoded and executed in the same way as every other instruction, which simplifies and speeds instruction decoding. Secondly, they simplify language compilation by providing a completely uniform way of allowing any instruction to take an operand of any size. Thirdly, they allow operands to be represented in a form independent of the processor wordlength.

3.2.3 Indirect functions

The remaining function code, *operate*, causes its operand to be interpreted as an operation on the values held in the evaluation stack. This allows up to 16 such operations to be encoded in a single byte instruction. However, the prefix instructions can be used to extend the operand of an *operate* instruction just like any other. The instruction representation therefore provides for an indefinite number of operations.

Encoding of the indirect functions is chosen so that the most frequently occurring operations are represented without the use of a prefix instruction. These include arithmetic, logical and comparison operations such as *add*, *exclusive or* and *greater than*. Less frequently occurring operations have encodings which require a single prefix operation.

3.2.4 Expression evaluation

Evaluation of expressions sometimes requires use of temporary variables in the workspace, but the number of these can be minimised by careful choice of the evaluation order.

Table 3.2: Expression evaluation

Program	Mnemonic	
x := 0	*ldc*	*0*
	stl	*x*
x := #24	*pfix*	*2*
	ldc	*4*
	stl	*x*
x := y + z	*ldl*	*y*
	ldl	*z*
	add	
	stl	*x*

3.2.5 Efficiency of encoding

Measurements show that about 70% of executed instructions are encoded in a single byte; that is, without the use of prefix instructions. Many of these instructions, such as *load constant* and *add* require just one processor cycle.

The instruction representation gives a more compact representation of high level language programs than more conventional instruction sets. Since a program requires less store to represent it, less of the memory bandwidth is taken up with fetching instructions. Furthermore, as memory is word accessed the processor will receive two instructions for every fetch.

Short instructions also improve the effectiveness of instruction pre-fetch, which in turn improves processor performance. There is an extra word of pre-fetch buffer, so the processor rarely has to wait for an instruction fetch before proceeding. Since the buffer is short, there is little time penalty when a jump instruction causes the buffer contents to be discarded.

3.3 Processes and concurrency

A process starts, performs a number of actions, and then either stops without completing or terminates complete. Typically, a process is a sequence of instructions. A transputer can run several processes in parallel (concurrently). Processes may be assigned either high or low priority, and there may be any number of each (page 173).

The processor has a microcoded scheduler which enables any number of concurrent processes to be executed together, sharing the processor time. This removes the need for a software kernel.

At any time, a concurrent process may be

Active	-	Being executed.
	-	On a list waiting to be executed.
Inactive	-	Ready to input.
	-	Ready to output.
	-	Waiting until a specified time.

The scheduler operates in such a way that inactive processes do not consume any processor time. It allocates a portion of the processor's time to each process in turn. Active processes waiting to be executed are held in two linked lists of process workspaces, one of high priority processes and one of low priority processes (page 173). Each list is implemented using two registers, one of which points to the first process in the list, the other to the last. In the Linked Process List figure 3.3, process *S* is executing and *P*, *Q* and *R* are active, awaiting execution. Only the low priority process queue registers are shown; the high priority process ones perform in a similar manner.

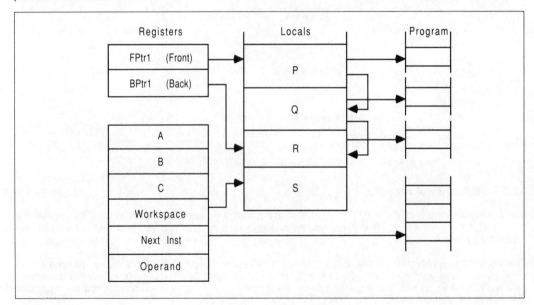

Figure 3.3: Linked process list

Table 3.3: Priority queue control registers

Function	High Priority	Low Priority
Pointer to front of active process list	*Fptr0*	*Fptr1*
Pointer to back of active process list	*Bptr0*	*Bptr1*

3   Processor

Each process runs until it has completed its action, but is descheduled whilst waiting for communication from another process or transputer, or for a time delay to complete. In order for several processes to operate in parallel, a low priority process is only permitted to run for a maximum of two time slices before it is forcibly descheduled at the next descheduling point (page 177). The time slice period is 5120 cycles of the external 5 MHz clock, giving ticks approximately 1ms apart.

A process can only be descheduled on certain instructions, known as descheduling points (page 177). As a result, an expression evaluation can be guaranteed to execute without the process being timesliced part way through.

Whenever a process is unable to proceed, its instruction pointer is saved in the process workspace and the next process taken from the list. Process scheduling pointers are updated by instructions which cause scheduling operations, and should not be altered directly. Actual process switch times are less than 1 μs, as little state needs to be saved and it is not necessary to save the evaluation stack on rescheduling.

The processor provides a number of special operations to support the process model, including *start process* and *end process*. When a main process executes a parallel construct, *start process* instructions are used to create the necessary additional concurrent processes. A *start process* instruction creates a new process by adding a new workspace to the end of the scheduling list, enabling the new concurrent process to be executed together with the ones already being executed. When a process is made active it is always added to the end of the list, and thus cannot pre-empt processes already on the same list.

The correct termination of a parallel construct is assured by use of the *end process* instruction. This uses a workspace location as a counter of the parallel construct components which have still to terminate. The counter is initialised to the number of components before the processes are *started*. Each component ends with an *end process* instruction which decrements and tests the counter. For all but the last component, the counter is non zero and the component is descheduled. For the last component, the counter is zero and the main process continues.

3.4   Priority

The IMS T212 supports two levels of priority. Priority 1 (low priority) processes are executed whenever there are no active priority 0 (high priority) processes.

High priority processes are expected to execute for a short time. If one or more high priority processes are able to proceed, then one is selected and runs until it has to wait for a communication, a timer input, or until it completes processing.

If no process at high priority is able to proceed, but one or more processes at low priority are able to proceed, then one is selected.

Low priority processes are periodically timesliced to provide an even distribution of processor time between computationally intensive tasks.

If there are **n** low priority processes, then the maximum latency from the time at which a low priority process becomes active to the time when it starts processing is 2**n**-2 timeslice periods. It is then able to execute for between one and two timeslice periods, less any time taken by high priority processes. This assumes that no process monopolises the transputer's time; i.e. it has a distribution of descheduling points (page 177).

Each timeslice period lasts for 5120 cycles of the external 5 MHz input clock (approximately 1 millisecond at the standard frequency of 5 MHz).

If a high priority process is waiting for an external channel to become ready, and if no other high priority process is active, then the interrupt latency (from when the channel becomes ready to when the process starts executing) is typically 19 processor cycles, a maximum of 53 cycles (assuming use of on-chip RAM).

3.5 Communications

Communication between processes is achieved by means of channels. Process communication is point-to-point, synchronised and unbuffered. As a result, a channel needs no process queue, no message queue and no message buffer.

A channel between two processes executing on the same transputer is implemented by a single word in memory; a channel between processes executing on different transputers is implemented by point-to-point links. The processor provides a number of operations to support message passing, the most important being *input message* and *output message*.

The *input message* and *output message* instructions use the address of the channel to determine whether the channel is internal or external. Thus the same instruction sequence can be used for both, allowing a process to be written and compiled without knowledge of where its channels are connected.

The process which first becomes ready must wait until the second one is also ready. A process performs an input or output by loading the evaluation stack with a pointer to a message, the address of a channel, and a count of the number of bytes to be transferred, and then executing an *input message* or *output message* instruction. Data is transferred if the other process is ready. If the channel is not ready or is an external one the process will deschedule.

3.6 Timers

The transputer has two 16 bit timer clocks which 'tick' periodically. The timers provide accurate process timing, allowing processes to deschedule themselves until a specific time.

One timer is accessible only to high priority processes and is incremented every microsecond, cycling completely in approximately 65 milliseconds. The other is accessible only to low priority processes and is incremented every 64 microseconds, giving exactly 15625 ticks in one second. It has a full period of approximately four seconds.

Table 3.4: Timer registers

Clock0	Current value of high priority (level 0) process clock
Clock1	Current value of low priority (level 1) process clock
TNextReg0	Indicates time of earliest event on high priority (level 0) timer queue
TNextReg1	Indicates time of earliest event on low priority (level 1) timer queue

The current value of the processor clock can be read by executing a *load timer* instruction. A process can arrange to perform a *timer input*, in which case it will become ready to execute after a specified time has been reached. The *timer input* instruction requires a time to be specified. If this time is in the 'past' then the instruction has no effect. If the time is in the 'future' then the process is descheduled. When the specified time is reached the process is scheduled again.

Figure 3.4 shows two processes waiting on the timer queue, one waiting for time 21, the other for time 31.

3 Processor

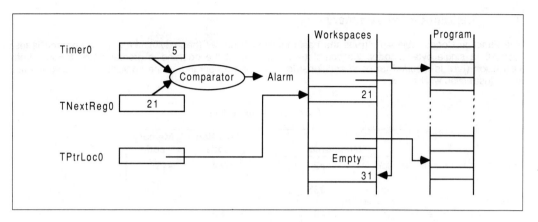

Figure 3.4: Timer registers

4   Instruction set summary

The Function Codes table 4.7. gives the basic function code set (page 170). Where the operand is less than 16, a single byte encodes the complete instruction. If the operand is greater than 15, one prefix instruction (*pfix*) is required for each additional four bits of the operand. If the operand is negative the first prefix instruction will be *nfix*.

Table 4.1: *prefix* coding

Mnemonic		Function code	Memory code
ldc	#3	#4	#43
ldc	#35		
is coded as			
pfix	#3	#2	#23
ldc	#5	#4	#45
ldc	#987		
is coded as			
pfix	#9	#2	#29
pfix	#8	#2	#28
ldc	#7	#4	#47
ldc	-31 (*ldc* #FFE1)		
is coded as			
nfix	#1	#6	#61
ldc	#1	#4	#41

Tables 4.8 to 4.17 give details of the operation codes. Where an operation code is less than 16 (e.g. *add*: operation code **05**), the operation can be stored as a single byte comprising the *operate* function code **F** and the operand (**5** in the example). Where an operation code is greater than 15 (e.g. *ladd*: operation code **16**), the *prefix* function code **2** is used to extend the instruction.

Table 4.2: *operate* coding

Mnemonic		Function code	Memory code
add	(op. code #5)		#F5
is coded as			
opr	add	#F	#F5
ladd	(op. code #16)		#21F6
is coded as			
pfix	#1	#2	#21
opr	#6	#F	#F6

4 Instruction set summary

The Processor Cycles column refers to the number of periods **TPCLPCL** taken by an instruction executing in internal memory. The number of cycles is given for the basic operation only; where relevant the time for the *prefix* function (one cycle) should be added. For a 20 MHz transputer one cycle is 50ns. Some instruction times vary. Where a letter is included in the cycles column it is interpreted from table 4.3.

Table 4.3: Instruction set interpretation

Ident	Interpretation
b	Bit number of the highest bit set in register A. Bit 0 is the least significant bit.
n	Number of places shifted.
w	Number of words in the message. Part words are counted as full words. If the message is not word aligned the number of words is increased to include the part words at either end of the message.

The **DE** column of the tables indicates the descheduling/error features of an instruction as described in table 4.4.

Table 4.4: Instruction features

Ident	Feature	See page:
D	The instruction is a descheduling point	177
E	The instruction will affect the *Error* flag	177, 186

4.1 Descheduling points

The instructions in table 4.5 are the only ones at which a process may be descheduled (page 172). They are also the ones at which the processor will halt if the **Analyse** pin is asserted (page 185).

Table 4.5: Descheduling point instructions

input message	output message	output byte	output word
timer alt wait	timer input	stop on error	alt wait
jump	loop end	end process	stop process

4.2 Error instructions

The instructions in table 4.6 are the only ones which can affect the *Error* flag (page 186) directly.

Table 4.6: Error setting instructions

add	add constant	subtract	
multiply		divide	remainder
long add	long subtract	long divide	
set error	testerr		
check word	check subscript from 0	check single	check count from 1

Table 4.7: IMS T212 function codes

Function Code	Memory Code	Mnemonic	Processor Cycles	Name	D E
0	0X	j	3	jump	D
1	1X	ldlp	1	load local pointer	
2	2X	pfix	1	prefix	
3	3X	ldnl	2	load non-local	
4	4X	ldc	1	load constant	
5	5X	ldnlp	1	load non-local pointer	
6	6X	nfix	1	negative prefix	
7	7X	ldl	2	load local	
8	8X	adc	1	add constant	E
9	9X	call	7	call	
A	AX	cj	2	conditional jump (not taken)	
			4	conditional jump (taken)	
B	BX	ajw	1	adjust workspace	
C	CX	eqc	2	equals constant	
D	DX	stl	1	store local	
E	EX	stnl	2	store non-local	
F	FX	opr	-	operate	

Table 4.8: IMS T212 arithmetic/logical operation codes

Operation Code	Memory Code	Mnemonic	Processor Cycles	Name	D E
46	24F6	and	1	and	
4B	24FB	or	1	or	
33	23F3	xor	1	exclusive or	
32	23F2	not	1	bitwise not	
41	24F1	shl	$n+2$	shift left	
40	24F0	shr	$n+2$	shift right	
05	F5	add	1	add	E
0C	FC	sub	1	subtract	E
53	25F3	mul	26	multiply	E
2C	22FC	div	23	divide	E
1F	21FF	rem	21	remainder	E
09	F9	gt	2	greater than	
04	F4	diff	1	difference	
52	25F2	sum	1	sum	
08	F8	prod	$b+4$	product	

4 Instruction set summary

Table 4.9: IMS T212 long arithmetic operation codes

Operation Code	Memory Code	Mnemonic	Processor Cycles	Name	DE
16	21F6	ladd	2	long add	E
38	23F8	lsub	2	long subtract	E
37	23F7	lsum	2	long sum	
4F	24FF	ldiff	2	long diff	
31	23F1	lmul	17	long multiply	
1A	21FA	ldiv	19	long divide	E
36	23F6	lshl	n+3	long shift left ($n<16$)	
			n-12	long shift left ($n \geq 16$)	
35	23F5	lshr	n+3	long shift right ($n<16$)	
			n-12	long shift right ($n \geq 16$)	
19	21F9	norm	n+5	normalise ($n<16$)	
			n-10	normalise ($n \geq 16$)	
			3	normalise ($n=32$)	

Table 4.10: IMS T212 general operation codes

Operation Code	Memory Code	Mnemonic	Processor Cycles	Name	DE
00	F0	rev	1	reverse	
3A	23FA	xword	4	extend to word	
56	25F6	cword	5	check word	E
1D	21FD	xdble	2	extend to double	
4C	24FC	csngl	3	check single	E
42	24F2	mint	1	minimum integer	

Table 4.11: IMS T212 indexing/array operation codes

Operation Code	Memory Code	Mnemonic	Processor Cycles	Name	D E
02	F2	bsub	1	byte subscript	
0A	FA	wsub	2	word subscript	
34	23F4	bcnt	2	byte count	
3F	23FF	wcnt	4	word count	
01	F1	lb	5	load byte	
3B	23FB	sb	4	store byte	
4A	24FA	move	2w+8	move message	

Table 4.12: IMS T212 timer handling operation codes

Operation Code	Memory Code	Mnemonic	Processor Cycles	Name	D E
22	22F2	ldtimer	2	load timer	
2B	22FB	tin	30	timer input (time future)	D
			4	timer input (time past)	D
4E	24FE	talt	4	timer alt start	
51	25F1	taltwt	15	timer alt wait (time past)	D
			48	timer alt wait (time future)	D
47	24F7	enbt	8	enable timer	
2E	22FE	dist	23	disable timer	

Table 4.13: IMS T212 input/output operation codes

Operation Code	Memory Code	Mnemonic	Processor Cycles	Name	D E
07	F7	in	2w+19	input message	D
0B	FB	out	2w+19	output message	D
0F	FF	outword	23	output word	D
0E	FE	outbyte	23	output byte	D
12	21F2	resetch	3	reset channel	
43	24F3	alt	2	alt start	
44	24F4	altwt	5	alt wait (channel ready)	D
			17	alt wait (channel not ready)	D
45	24F5	altend	4	alt end	
49	24F9	enbs	3	enable skip	
30	23F0	diss	4	disable skip	
48	24F8	enbc	7	enable channel (ready)	
			5	enable channel (not ready)	
2F	22FF	disc	8	disable channel	

4 Instruction set summary

Table 4.14: IMS T212 control operation codes

Operation Code	Memory Code	Mnemonic	Processor Cycles	Name	DE
20	22F0	ret	5	return	
1B	21FB	ldpi	2	load pointer to instruction	
3C	23FC	gajw	2	general adjust workspace	
06	F6	gcall	4	general call	
21	22F1	lend	10	loop end (loop)	D
			5	loop end (exit)	D

Table 4.15: IMS T212 scheduling operation codes

Operation Code	Memory Code	Mnemonic	Processor Cycles	Name	DE
0D	FD	startp	12	start process	D
03	F3	endp	13	end process	D
39	23F9	runp	10	run process	
15	21F5	stopp	11	stop process	
1E	21FE	ldpri	1	load current priority	

Table 4.16: IMS T212 error handling operation codes

Operation Code	Memory Code	Mnemonic	Processor Cycles	Name	DE
13	21F3	csub0	2	check subscript from 0	E
4D	24FD	ccnt1	3	check count from 1	E
29	22F9	testerr	2	test error false and clear (no error)	
			3	test error false and clear (error)	
10	21F0	seterr	1	set error	E
55	25F5	stoperr	2	stop on error (no error)	D
57	25F7	clrhalterr	1	clear halt-on-error	
58	25F8	sethalterr	1	set halt-on-error	
59	25F9	testhalterr	2	test halt-on-error	

Table 4.17: IMS T212 processor initialisation operation codes

Operation Code	Memory Code	Mnemonic	Processor Cycles	Name	DE
2A	22FA	testpranal	2	test processor analysing	
3E	23FE	saveh	4	save high priority queue registers	
3D	23FD	savel	4	save low priority queue registers	
18	21F8	sthf	1	store high priority front pointer	
50	25F0	sthb	1	store high priority back pointer	
1C	21FC	stlf	1	store low priority front pointer	
17	21F7	stlb	1	store low priority back pointer	
54	25F4	sttimer	1	store timer	

5 System services

System services include all the necessary logic to initialise and sustain operation of the device. They also include error handling and analysis facilities.

5.1 Power

Power is supplied to the device via the **VCC** and **GND** pins. Several of each are provided to minimise inductance within the package. All supply pins must be connected. The supply must be decoupled close to the chip by at least one 100nF low inductance (e.g. ceramic) capacitor between **VCC** and **GND**. Four layer boards are recommended; if two layer boards are used, extra care should be taken in decoupling.

Input voltages must not exceed specification with respect to **VCC** and **GND**, even during power-up and power-down ramping, otherwise *latchup* can occur. CMOS devices can be permanently damaged by excessive periods of latchup.

5.2 CapPlus, CapMinus

The internally derived power supply for internal clocks requires an external low leakage, low inductance $1\mu F$ capacitor to be connected between **CapPlus** and **CapMinus**. A ceramic capacitor is preferred, with an impedance less than 3 ohms between 100 KHz and 20 MHz. If a polarised capacitor is used the negative terminal should be connected to **CapMinus**. Total PCB track length should be less than 50mm. The connections must not touch power supplies or other noise sources.

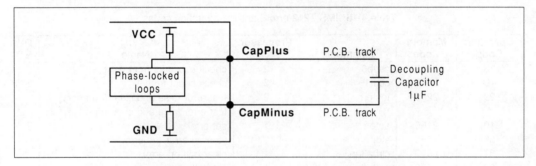

Figure 5.1: Recommended PLL decoupling

5.3 ClockIn

Transputer family components use a standard clock frequency, supplied by the user on the **ClockIn** input. The nominal frequency of this clock for all transputer family components is 5MHz, regardless of device type, transputer word length or processor cycle time. High frequency internal clocks are derived from **ClockIn**, simplifying system design and avoiding problems of distributing high speed clocks externally.

A number of transputer devices may be connected to a common clock, or may have individual clocks providing each one meets the specified stability criteria. In a multi-clock system the relative phasing of **ClockIn** clocks is not important, due to the asynchronous nature of the links. Mark/space ratio is unimportant provided the specified limits of **ClockIn** pulse widths are met.

Oscillator stability is important. **ClockIn** must be derived from a crystal oscillator; RC oscillators are not sufficiently stable. **ClockIn** must not be distributed through a long chain of buffers. Clock edges must be monotonic and remain within the specified voltage and time limits.

5 System services

Table 5.1: Input clock

SYMBOL	PARAMETER	MIN	NOM	MAX	UNITS	NOTE
TDCLDCH	ClockIn pulse width low	40			ns	
TDCHDCL	ClockIn pulse width high	40			ns	
TDCLDCL	ClockIn period		200		ns	1,3
TDCerror	ClockIn timing error			±0.5	ns	2
TDC1DC2	Difference in ClockIn for 2 linked devices			400	ppm	3
TDCr	ClockIn rise time			10	ns	4
TDCf	ClockIn fall time			8	ns	4

Notes

1 Measured between corresponding points on consecutive falling edges.

2 Variation of individual falling edges from their nominal times.

3 This value allows the use of 200ppm crystal oscillators for two devices connected together by a link.

4 Clock transitions must be monotonic within the range **VIH** to **VIL** (page 204).

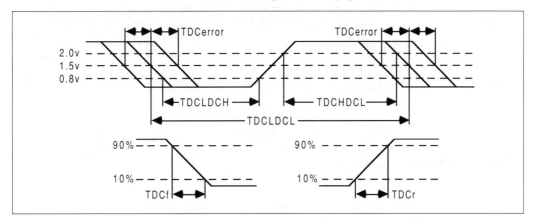

Figure 5.2: ClockIn timing

5.4 Reset

Reset can go high with **VCC**, but must at no time exceed the maximum specified voltage for **VIH**. After **VCC** is valid **ClockIn** should be running for a minimum period **TDCVRL** before the end of **Reset**. The falling edge of **Reset** initialises the transputer and starts the bootstrap routine. Link outputs are forced low during reset; link inputs and **EventReq** should be held low. Memory request (DMA) must not occur whilst **Reset** is high but can occur before bootstrap (page 198). If **BootFromRom** is high bootstrapping will take place immediately after **Reset** goes low, using data from external memory; otherwise the transputer will await an input from any link. The processor will be in the low priority state.

5.5 Bootstrap

The transputer can be bootstrapped either from a link or from external ROM. To facilitate debugging, **BootFromRom** may be dynamically changed but must obey the specified timing restrictions.

If **BootFromRom** is connected high (e.g. to **VCC**) the transputer starts to execute code from the top two bytes in external memory, at address #7FFE. This location should contain a backward jump to a program in ROM. The processor is in the low priority state, and the *W* register points to *MemStart* (page 187).

Table 5.2: Reset and Analyse

SYMBOL	PARAMETER	MIN	NOM	MAX	UNITS	NOTE
TPVRH	Power valid before Reset	10			ms	
TRHRL	Reset pulse width high	8			ClockIn	1
TDCVRL	ClockIn running before Reset end	10			ms	2
TAHRH	Analyse setup before Reset	3			ms	
TRLAL	Analyse hold after Reset end	1			ns	
TBRVRL	BootFromRom setup	0			ms	
TRLBRX	BootFromRom hold after Reset	50			ms	
TALBRX	BootFromRom hold after Analyse	50			ms	

Notes

1 Full periods of **ClockIn TDCLDCL** required.

2 At power-on reset.

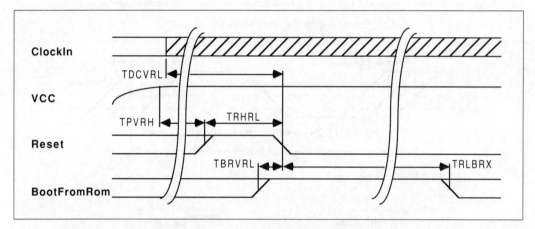

Figure 5.3: Transputer reset timing with Analyse low

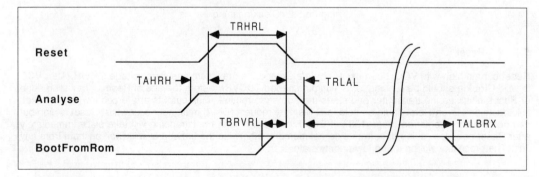

Figure 5.4: Transputer reset and analyse timing

5   System services

If **BootFromRom** is connected low (e.g. to **GND**) the transputer will wait for the first bootstrap message to arrive on any one of its links. The transputer is ready to receive the first byte on a link within two processor cycles **TPCLPCL** after **Reset** goes low.

If the first byte received (the control byte) is greater than 1 it is taken as the quantity of bytes to be input. The following bytes, to that quantity, are then placed in internal memory starting at location *MemStart*. Following reception of the last byte the transputer will start executing code at *MemStart* as a low priority process. The memory space immediately above the loaded code is used as work space. Messages arriving on other links after the control byte has been received and on the bootstrapping link after the last bootstrap byte will be retained until a process inputs from them.

5.6   Peek and poke

Any location in internal or external memory can be interrogated and altered when the transputer is waiting for a bootstrap from link. If the control byte is 0 then four more bytes are expected on the same link. The first two byte word is taken as an internal or external memory address at which to poke (write) the second two byte word. If the control byte is 1 the next two bytes are used as the address from which to peek (read) a word of data; the word is sent down the output channel of the same link.

Following such a peek or poke, the transputer returns to its previously held state. Any number of accesses may be made in this way until the control byte is greater than 1, when the transputer will commence reading its bootstrap program. Any link can be used, but addresses and data must be transmitted via the same link as the control byte.

5.7   Analyse

If **Analyse** is taken high when the transputer is running, the transputer will halt at the next descheduling point (page 177). From **Analyse** being asserted, the processor will halt within three time slice periods plus the time taken for any high priority process to complete. As much of the transputer status is maintained as is necessary to permit analysis of the halted machine.

Input links will continue with outstanding transfers. Output links will not make another access to memory for data but will transmit only those bytes already in the link buffer. Providing there is no delay in link acknowledgement, the links should be inactive within a few microseconds of the transputer halting.

Reset should not be asserted before the transputer has halted and link transfers have ceased. If **BootFromRom** is high the transputer will bootstrap as soon as **Analyse** is taken low, otherwise it will await a control byte on any link. If **Analyse** is taken low without **Reset** going high the transputer state and operation are undefined. After the end of a valid **Analyse** sequence the registers have the values given in table 5.3.

Table 5.3: Register values after analyse

I	*MemStart* if bootstrapping from a link, or the external memory bootstrap address if bootstrapping from ROM.
W	*MemStart* if bootstrapping from ROM, or the address of the first free word after the bootstrap program if bootstrapping from link.
A	The value of *I* when the processor halted.
B	The value of *W* when the processor halted, together with the priority of the process when the transputer was halted (i.e. the *W* descriptor).
C	The ID of the bootstrapping link if bootstrapping from link.

5.8   Error

The **Error** pin is connected directly to the internal *Error* flag and follows the state of that flag. If **Error** is high it indicates an error in one of the processes caused, for example, by arithmetic overflow, divide by zero, array bounds violation or software setting the flag directly (page 177). Once set, the *Error* flag is only cleared by executing the instruction *testerr* (page 176). The error is not cleared by processor reset, in order that analysis can identify any errant transputer (page 185).

A process can be programmed to stop if the *Error* flag is set; it cannot then transmit erroneous data to other processes, but processes which do not require that data can still be scheduled. Eventually all processes which rely, directly or indirectly, on data from the process in error will stop through lack of data.

By setting the *HaltOnError* flag the transputer itself can be programmed to halt if *Error* becomes set. If *Error* becomes set after *HaltOnError* has been set, all processes on that transputer will cease but will not necessarily cause other transputers in a network to halt. Setting *HaltOnError* after *Error* will not cause the transputer to halt; this allows the processor reset and analyse facilities to function with the flags in indeterminate states.

An alternative method of error handling is to have the errant process or transputer cause all transputers to halt. This can be done by applying the **Error** output signal of the errant transputer to the **EventReq** pin of a suitably programmed master transputer. Since the process state is preserved when stopped by an error, the master transputer can then use the analyse function to debug the fault. When using such a circuit, note that the *Error* flag is in an indeterminate state on power up; the circuit and software should be designed with this in mind.

Error checks can be removed completely to optimise the performance of a proven program; any unexpected error then occurring will have an arbitrary undefined effect.

If a high priority process pre-empts a low priority one, status of the *Error* and *HaltOnError* flags is saved for the duration of the high priority process and restored at the conclusion of it. Status of the *Error* flag is transmitted to the high priority process but the *HaltOnError* flag is cleared before the process starts. Either flag can be altered in the process without upsetting the error status of any complex operation being carried out by the pre-empted low priority process.

In the event of a transputer halting because of *HaltOnError*, the links will finish outstanding transfers before shutting down. If **Analyse** is asserted then all inputs continue but outputs will not make another access to memory for data.

After halting due to the *Error* flag changing from 0 to 1 whilst *HaltOnError* is set, register *I* points two bytes past the instruction which set *Error*. After halting due to the **Analyse** pin being taken high, register *I* points one byte past the instruction being executed. In both cases *I* will be copied to register *A*.

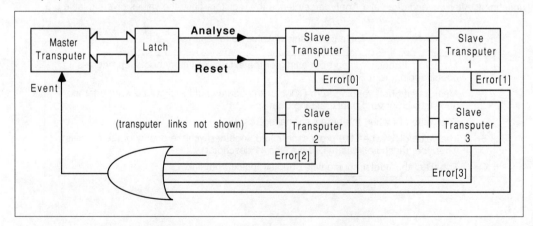

Figure 5.5: Error handling in a multi-transputer system

6 Memory

The IMS T212 has 2 Kbytes of fast internal static memory for high rates of data throughput. Each internal memory access takes one processor cycle **ProcClockOut** (page 189). The transputer can also access an additional 62 Kbytes of external memory space. Internal and external memory are part of the same linear address space.

IMS T212 memory is byte addressed, with words aligned on two-byte boundaries. The least significant byte of a word is the lowest addressed byte.

The bits in a byte are numbered 0 to 7, with bit 0 the least significant. The bytes are numbered from 0, with byte 0 the least significant. In general, wherever a value is treated as a number of component values, the components are numbered in order of increasing numerical significance, with the least significant component numbered 0. Where values are stored in memory, the least significant component value is stored at the lowest (most negative) address.

Internal memory starts at the most negative address #8000 and extends to #87FF. User memory begins at #8024; this location is given the name *MemStart*.

A reserved area at the bottom of internal memory is used to implement link and event channels.

Two words of memory are reserved for timer use, *TPtrLoc0* for high priority processes and *TPtrLoc1* for low priority processes. They either indicate the relevant priority timer is not in use or point to the first process on the timer queue at that priority level.

Values of certain processor registers for the current low priority process are saved in the reserved *IntSaveLoc* locations when a high priority process pre-empts a low priority one.

External memory space starts at #8800 and extends up through #0000 to #7FFF. ROM bootstrapping code must be in the most positive address space, starting at #7FFE. Address space immediately below this is conventionally used for ROM based code.

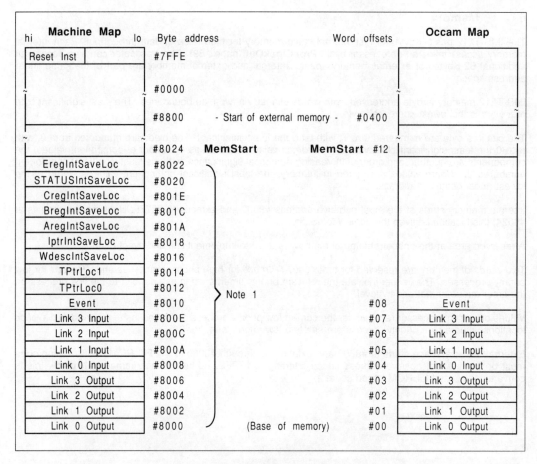

Figure 6.1: IMS T212 memory map

These locations are used as auxiliary processor registers and should not be manipulated by the user. Like processor registers, their contents may be useful for implementing debugging tools (**Analyse**, page 185). For details see The Transputer Instruction Set - A Compiler Writers' Guide.

7 External memory interface

The IMS T212 External Memory Interface (EMI) allows access to a 16 bit address space via separate address and data buses. The data bus can be configured for either 16 bit or 8 bit memory access, allowing the use of a single bank of byte-wide memory. Both word-wide and byte-wide access may be mixed in a single memory system (page 195).

7.1 ProcClockOut

This clock is derived from the internal processor clock, which is in turn derived from **ClockIn**. Its period is equal to one internal microcode cycle time, and can be derived from the formula

$$TPCLPCL = TDCLDCL / PLLx$$

where **TPCLPCL** is the **ProcClockOut Period**, **TDCLDCL** is the **ClockIn Period** and **PLLx** is the phase lock loop factor for the relevant speed part, obtained from the ordering details (Ordering appendix).

Edges of the various external memory strobes are synchronised by, but do not all coincide with, rising or falling edges of **ProcClockOut**.

Table 7.1: ProcClockOut

SYMBOL	PARAMETER	MIN	NOM	MAX	UNITS	NOTE
TPCLPCL	ProcClockOut period	a-1	a	a+1	ns	1
TPCHPCL	ProcClockOut pulse width high	b-2.5	b	b+2.5	ns	2
TPCLPCH	ProcClockOut pulse width low		c		ns	3
TPCstab	ProcClockOut stability			4	%	4

Notes

1 a is **TDCLDCL/PLLx**.

2 b is 0.5∗**TPCLPCL** (half the processor clock period).

3 c is **TPCLPCL-TPCHPCL**.

4 Stability is the variation of cycle periods between two consecutive cycles, measured at corresponding points on the cycles.

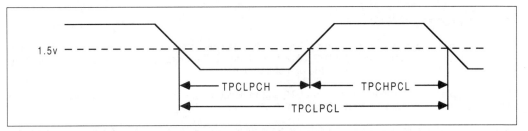

Figure 7.1: IMS T212 **ProcClockOut** timing

7.2 Tstates

The external memory cycle is divided into four **Tstates** with the following functions:

- **T1** Address and control setup time.
- **T2** Data setup time.
- **T3** Data read/write.
- **T4** Data and address hold after access.

Each **Tstate** is half a processor cycle **TPCLPCL** long, displaced by approximately one fourth of a cycle from **ProcClockOut** edges. **T2** can be extended indefinitely by adding externally generated wait states of one complete processor cycle each.

An external memory cycle is always a complete number of cycles **TPCLPCL** in length. The start of **T1** always coincides with the low phase of **ProcClockOut**.

7.3 Internal access

During an internal memory access cycle the external memory interface address bus **MemA0-15** reflects the word address used to access internal RAM, **notMemWrB0-1** reflect the internal read/write operation, **notMemCE** is inactive and the data bus **MemD0-15** is tristated. This is true unless and until a DMA (memory request) activity takes place, when the lines will be placed in a high impedance state by the transputer.

Bus activity is not adequate to trace the internal operation of the transputer in full, but may be used for hardware debugging in conjuction with peek and poke (page 185).

7.4 MemA0-15

External memory addresses are output on a non-multiplexed 16 bit bus. The address is valid at the start of **T1** and remains so until the end of **T4**, with the timing shown. Byte addressing is carried out internally by the IMS T212 for read cycles. For write cycles the relevant bytes in memory are addressed by the write enables **notMemWrB0-1**.

The transputer places the address bus in a high impedance state during DMA.

7.5 MemD0-15

The non-multiplexed data bus is 16 bits wide. Read cycle data may be set up on the bus at any time after the start of **T1**, but must be valid when the IMS T212 reads it during **T3**. Data can be removed any time during **T4**, but must be off the bus no later than the end of that period.

Write data is placed on the bus at the start of **T2** and removed at the end of **T4**. It is normally written into memory in synchronism with **notMemCE** going high.

The data bus is high impedance except when the transputer is writing data. If only one byte is being written, the unused 8 bits of the bus are high impedance at that time. In byte access mode **MemD8-15** are high impedance during the external memory cycle which writes the most significant (second) byte (page 195).

If the data setup time for read or write is too short it can be extended by inserting wait states at the end of **T2** (page 196).

7 External memory interface

Table 7.2: Read

SYMBOL	PARAMETER	T212-20 MIN	T212-20 MAX	T212-17 MIN	T212-17 MAX	UNITS	NOTE
TAVEL	Address valid before chip enable low	13	16	15	19	ns	
TELEH	Chip enable low	56	63	65	72	ns	
TEHEL	Delay before chip enable re-assertion	35	46	40	51	ns	1
TEHAX	Address hold after chip enable high	20	24	21	27	ns	
TELDrV	Data valid from chip enable low		40		43	ns	
TDrVEH	Data setup before chip enable high	11		15		ns	
TEHDrZ	Data hold after chip enable high	0		0		ns	
TWEHEL	Write enable setup before chip enable low	14		18		ns	2

Notes

1 These values assume back-to-back external memory accesses.

2 Timing is for both write enables **notMemWrB0-1**.

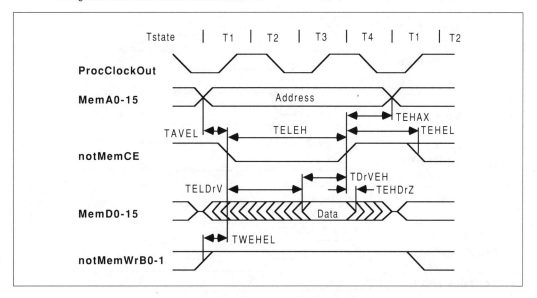

Figure 7.2: IMS T212 external read cycle

7.6 notMemWrB0-1

Two write enables are provided, one to write each byte of the word. When writing a word, both write enables are asserted; when writing a byte only the appropriate write enable is asserted. **notMemWrB0** addresses the least significant byte. The write enables are active before the chip enable signal **notMemCE** becomes active, thus reducing memory access time and the risk of bus contention.

Data must be strobed into memory by, or in conjunction with, **notMemCE**, as the write enables are not guaranteed to go high between consecutive write cycles. The write enables are placed in a high impedance state during DMA.

Table 7.3: Write

SYMBOL	PARAMETER	T212-20 MIN	T212-20 MAX	T212-17 MIN	T212-17 MAX	UNITS	NOTE
TDwVEH	Data setup before chip enable high	36		42		ns	
TEHDwZ	Data hold after write	22	30	24	32	ns	
TWELEL	Write enable setup before chip enable low	4	20	4	24	ns	1
TEHWEH	Write enable hold after chip enable high	17	25	18	27	ns	1

Notes

1 Timing is for both write enables **notMemWrB0-1**.

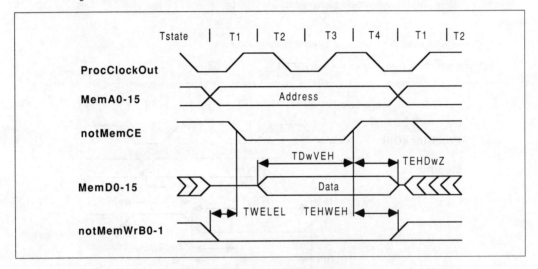

Figure 7.3: IMS T212 external write cycle

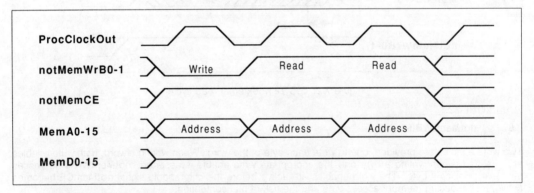

Figure 7.4: IMS T212 typical bus activity for internal memory cycles

7 External memory interface

7.7 notMemCE

The active low signal **notMemCE** is used to enable external memory on both read and write cycles. It must be used, in conjunction with the write enables **notMemWrB0-1**, to write data into memory; the write enable lines only select the byte of memory to be written.

Table 7.4: **notMemCE** to **ProcClockOut** skew

SYMBOL	PARAMETER	T212-20 MIN	T212-20 MAX	T212-17 MIN	T212-17 MAX	UNITS	NOTE
TPCHEL	notMemCe falling from ProcClockOut rising	1	5	2	8	ns	
TEHPCL	notMemCe rising to ProcClockOut falling	8	14	10	15	ns	

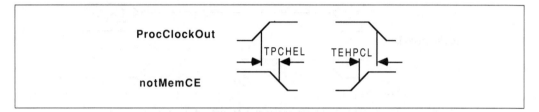

Figure 7.5: IMS T212 skew of **notMemCE** to **ProcClockOut**

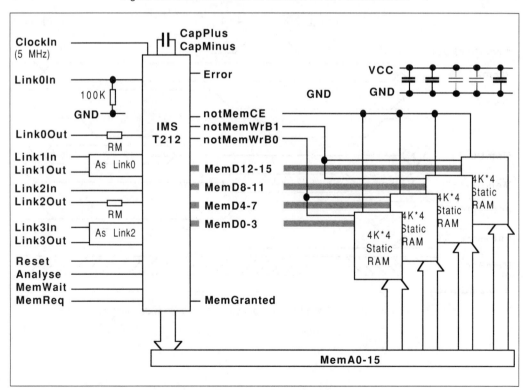

Figure 7.6: IMS T212 application

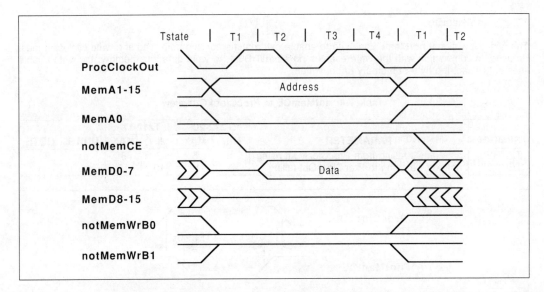

Figure 7.7: IMS T212 Least significant byte write in word access mode

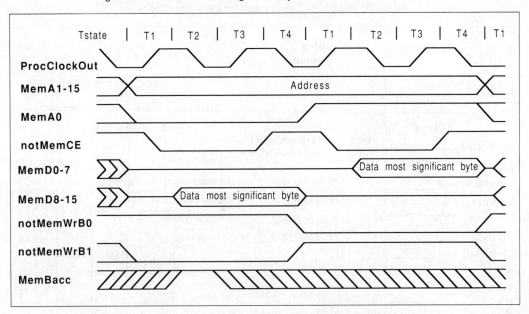

Figure 7.8: IMS T212 Most significant byte write to byte-wide memory

7 External memory interface

7.8 MemBAcc

The IMS T212 will, by default, perform word access at even memory locations. Access to byte-wide memory can be achieved by taking **MemBAcc** high with the timing shown. Where all external memory operations are to byte-wide memory, **MemBAcc** may be wired permanently high. The state of this signal is latched during **T2**.

If **MemBAcc** is low then a full word will be accessed in one external memory cycle, otherwise the high and low bytes of the word will be separately accessed during two consecutive cycles. The first (least significant) byte is accessed at the word address (**MemA0** is low). The second (most significant) byte is accessed at the word address +1 (**MemA0** is high).

With **MemBAcc** high, the first cycle is identical with a normal word access cycle. However, it will be immediately followed by another memory cycle, which will use **MemD0-7** to read or write the second (most significant) byte of data. During this second cycle **notMemWrB1** remains high, both for read and write, and **MemD8-15** are high impedance. When writing a single byte with **MemBAcc** high, both the first and second cycles are performed with **notMemWrB0** asserted in the appropriate cycle.

Table 7.5: Byte-wide memory access

SYMBOL	PARAMETER	T212-20 MIN	T212-20 MAX	T212-17 MIN	T212-17 MAX	UNITS	NOTE
TELBAH	MemBAcc high from chip enable		12		15	ns	
TELBAL	MemBAcc low from chip enable	26		29		ns	

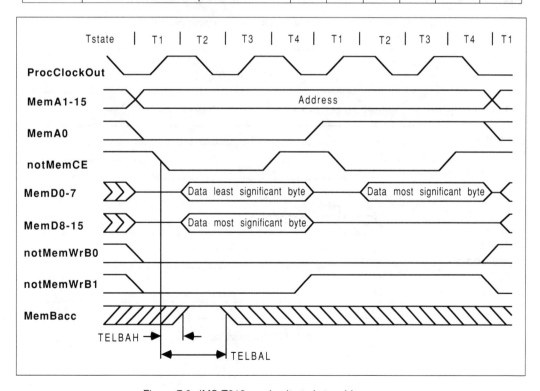

Figure 7.9: IMS T212 word write to byte-wide memory

7.9 MemWait

Taking **MemWait** high with the timing shown in the diagram will extend the duration of **T2** by one processor cycle **TPCLPCL**. One wait state comprises the pair **W1** and **W2**. **MemWait** is sampled near the falling edge of **ProcClockOut** during **T2**, and should not change state in this region. If **MemWait** is still high when sampled near the falling edge of **ProcClockOut** in **W2** then another wait period will be inserted. This can continue indefinitely.

The wait state generator can be a simple digital delay line, synchronised to **notMemCE**. The **Single Wait State Generator** circuit in figure 7.11 can be extended to provide two or more wait states, as shown in figure 7.12.

The **Programmable Wait State Generator** circuit in figure 7.13 is designed to be interfaced directly to any memory or peripheral enable signal; 'F' series devices should be employed to ensure minimum delay between **notMemCE** and a valid **notWaitX** input. Only one wait select input line should be low at any one time; for zero wait states **notWait0** must be asserted.

Table 7.6: Memory wait

SYMBOL	PARAMETER	T212-20 MIN	T212-20 MAX	T212-17 MIN	T212-17 MAX	UNITS	NOTE
TELWtH	MemWait asserted after chip enable low		13		13	ns	
TELWtL	Wait hold after chip enable low	23	a+13	23	a+13	ns	1

Notes

1. **a** is **w∗c** where **w** is the number of wait states and **c** is the toleranced clock period of 49 ns for IMS T212-20, 56 ns for IMS T212-17.

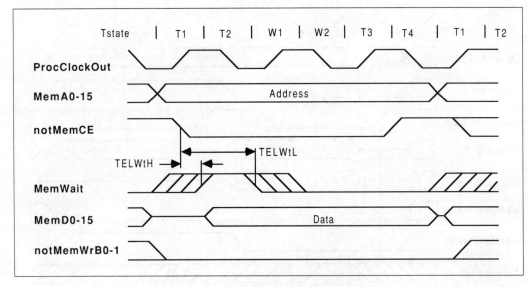

Figure 7.10: IMS T212 memory wait timing

7 External memory interface

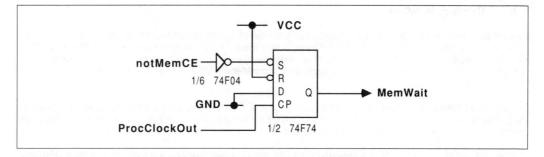

Figure 7.11: Single wait state generator

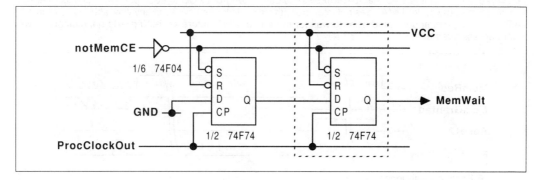

Figure 7.12: Extendable wait state generator

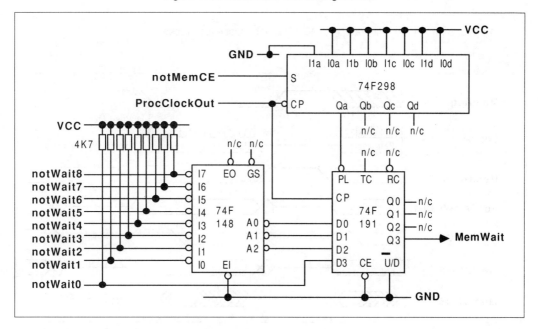

Figure 7.13: Programmable wait state generator

7.10 MemReq, MemGranted

Direct memory access (DMA) can be requested at any time by taking the asynchronous **MemReq** input high. For external memory cycles, the IMS T212 samples **MemReq** during the first high phase of **ProcClockOut** after **notMemCE** goes low. In the absence of an external memory cycle, **MemReq** is sampled during every high phase of **ProcClockOut**. **MemA0-15**, **MemD0-15**, **notMemWrB0-1** and **notMemCE** are tristated before **MemGranted** is asserted.

Removal of **MemReq** is sampled during each high phase of **ProcClockOut** and **MemGranted** removed with the timing shown. Further external bus activity, either external cycles or reflection of internal cycles, will commence during the next low phase of **ProcClockOut**.

Chip enable, write enables, address bus and data bus are in a high impedance state during DMA. External circuitry must ensure that **notMemCE** and **notMemWrB0-1** do not become active whilst control is being transferred; it is recommended that a 10K resistor is connected from **VCC** to each pin. DMA cannot interrupt an external memory cycle. DMA does not interfere with internal memory cycles in any way, although a program running in internal memory would have to wait for the end of DMA before accessing external memory. DMA cannot access internal memory.

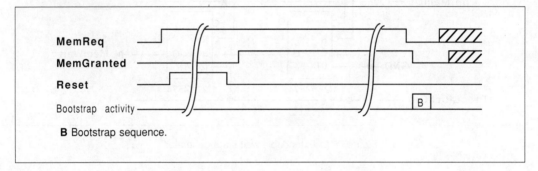

Figure 7.14: IMS T212 DMA sequence at reset

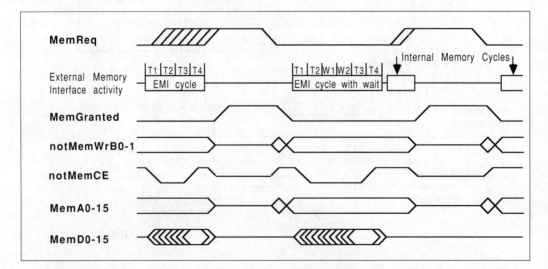

Figure 7.15: IMS T212 operation of **MemReq** and **MemGranted** with external and internal memory cycles

7 External memory interface

DMA allows a bootstrap program to be loaded into external RAM ready for execution after reset. If **MemReq** is held high throughout reset, **MemGranted** will be asserted before the bootstrap sequence begins. **MemReq** must be high at least one period **TDCLDCL** of **ClockIn** before **Reset**. The circuit should be designed to ensure correct operation if **Reset** could interrupt a normal DMA cycle.

Table 7.7: Memory request

SYMBOL	PARAMETER	T212-20 MIN	T212-20 MAX	T212-17 MIN	T212-17 MAX	UNITS	NOTE
TMRHMGH	Memory request response time	85	a	100	a	ns	1
TMRLMGL	Memory request end response time	90	100	100	114	ns	
TAZMGH	Addr. bus tristate before MemGranted	0		0		ns	
TAVMGL	Addr. bus active after MemGranted end	0		0		ns	
TDZMGH	Data bus tristate before MemGranted	0		0		ns	
TEZMGH	notMemCE tristate before MemGranted	0		0		ns	
TEVMGL	notMemCE active after MemGranted end	0		0		ns	
TWEZMGH	Write enable tristate before MemGranted	0		0		ns	
TWEVMGL	Write enable active after MemGranted end	0		0		ns	

Notes

1. Maximum response time **a** depends on whether an external memory cycle is in progress and whether byte access is active. Maximum time is (2 processor cycles) + (number of wait state cycles) for word access; in byte access mode this time is doubled.

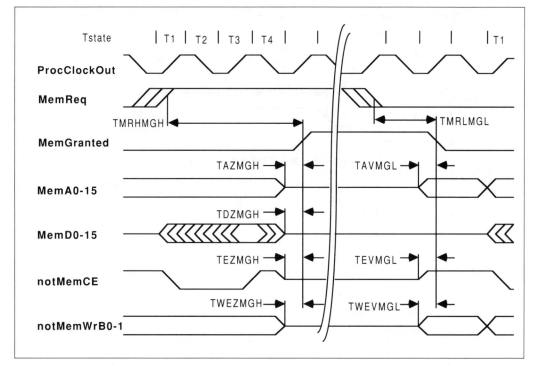

Figure 7.16: IMS T212 memory request timing

8 Events

EventReq and **EventAck** provide an asynchronous handshake interface between an external event and an internal process. When an external event takes **EventReq** high the external event channel (additional to the external link channels) is made ready to communicate with a process. When both the event channel and the process are ready the processor takes **EventAck** high and the process, if waiting, is scheduled. **EventAck** is removed after **EventReq** goes low.

Only one process may use the event channel at any given time. If no process requires an event to occur **EventAck** will never be taken high. Although **EventReq** triggers the channel on a transition from low to high, it must not be removed before **EventAck** is high. **EventReq** should be low during **Reset**; if not it will be ignored until it has gone low and returned high. **EventAck** is taken low when **Reset** occurs.

If the process is a high priority one and no other high priority process is running, the latency is as described on page 173. Setting a high priority task to wait for an event input is a way of interrupting a transputer program.

Table 8.1: Event

SYMBOL	PARAMETER	MIN	NOM	MAX	UNITS	NOTE
TVHKH	Event request response	0			ns	
TKHVL	Event request hold	0			ns	
TVLKL	Delay before removal of event acknowledge	0		a	ns	1
TKLVH	Delay before re-assertion of event request	0			ns	

Notes

1 **a** is **TPCLPCL** (2 periods **Tm**).

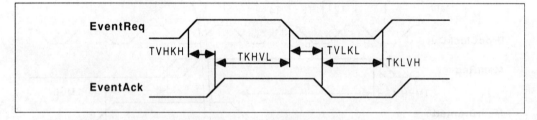

Figure 8.1: IMS T212 event timing

9 Links

Four identical INMOS bi-directional serial links provide synchronized communication between processors and with the outside world. Each link comprises an input channel and output channel. A link between two transputers is implemented by connecting a link interface on one transputer to a link interface on the other transputer Every byte of data sent on a link is acknowledged on the input of the same link, thus each signal line carries both data and control information.

The quiescent state of a link output is low. Each data byte is transmitted as a high start bit followed by a one bit followed by eight data bits followed by a low stop bit. The least significant bit of data is transmitted first. After transmitting a data byte the sender waits for the acknowledge, which consists of a high start bit followed by a zero bit. The acknowledge signifies both that a process was able to receive the acknowledged data byte and that the receiving link is able to receive another byte. The sending link reschedules the sending process only after the acknowledge for the final byte of the message has been received.

The IMS T212 links support the standard INMOS communication speed of 10 Mbits per second. In addition they can be used at 5 or 20 Mbits per second. Links are not synchronised with **ClockIn** or **ProcClockOut** and are insensitive to their phases. Thus links from independently clocked systems may communicate, providing only that the clocks are nominally identical and within specification.

Links are TTL compatible and intended to be used in electrically quiet environments, between devices on a single printed circuit board or between two boards via a backplane. Direct connection may be made between devices separated by a distance of less than 300 millimetres. For longer distances a matched 100 Ohm transmission line should be used with series matching resistors **RM**. When this is done the line delay should be less than 0.4 bit time to ensure that the reflection returns before the next data bit is sent.

Buffers may be used for very long transmissions. If so, their overall propagation delay should be stable within the skew tolerance of the link, although the absolute value of the delay is immaterial.

Link speeds can be set by **LinkSpecial**, **Link0Special** and **Link123Special**. The link 0 speed can be set independently. Table 9.1 shows uni-directional and bi-directional data rates in Kbytes/second for each link speed; **LinknSpecial** is to be read as **Link0Special** when selecting link 0 speed and as **Link123Special** for the others. Data rates are quoted for a transputer using internal memory, and will be affected by a factor depending on the number of external memory accesses and the length of the external memory cycle.

Table 9.1: Speed Settings for Transputer Links

Link Special	Linkn Special	Mbits/sec	Kbytes/sec	
			Uni	Bi
0	0	10	400	800
0	1	5	200	400
1	0	10	400	800
1	1	20	800	1600

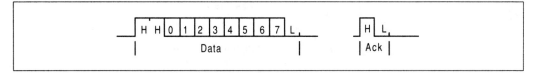

Figure 9.1: IMS T212 link data and acknowledge packets

Table 9.2: Link

SYMBOL	PARAMETER		MIN	NOM	MAX	UNITS	NOTE
TJQr	LinkOut rise time				20	ns	
TJQf	LinkOut fall time				10	ns	
TJDr	LinkIn rise time				20	ns	
TJDf	LinkIn fall time				20	ns	
TJQJD	Buffered edge delay		0			ns	
TJBskew	Variation in TJQJD	20 Mbits/s			3	ns	1
		10 Mbits/s			10	ns	1
		5 Mbits/s			30	ns	1
CLIZ	LinkIn capacitance	@ f=1MHz			7	pF	
CLL	LinkOut load capacitance				50	pF	
RM	Series resistor for 100Ω transmission line			56		ohms	

Notes

1 This is the variation in the total delay through buffers, transmission lines, differential receivers etc., caused by such things as short term variation in supply voltages and differences in delays for rising and falling edges.

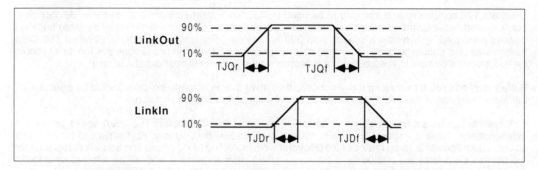

Figure 9.2: IMS T212 link timing

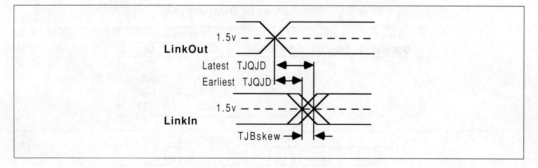

Figure 9.3: IMS T212 buffered link timing

9 Links

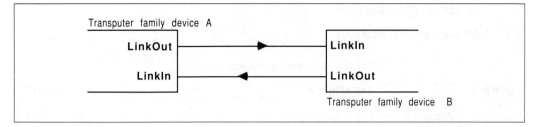

Figure 9.4: Links directly connected

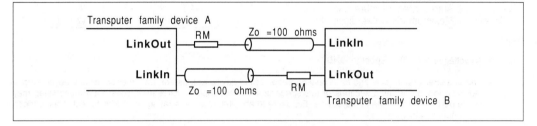

Figure 9.5: Links connected by transmission line

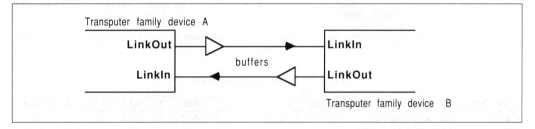

Figure 9.6: Links connected by buffers

10 Electrical specifications

10.1 DC electrical characteristics

Table 10.1: Absolute maximum ratings

SYMBOL	PARAMETER	MIN	MAX	UNITS	NOTE
VCC	DC supply voltage	0	7.0	V	1,2,3
VI, VO	Voltage on input and output pins	-0.5	VCC+0.5	V	1,2,3
II	Input current		±25	mA	4
OSCT	Output short circuit time (one pin)		1	s	2
TS	Storage temperature	-65	150	°C	2
TA	Ambient temperature under bias	-55	125	°C	2
PDmax	Maximum allowable dissipation		2	W	

Notes

1. All voltages are with respect to **GND**.

2. This is a stress rating only and functional operation of the device at these or any other conditions beyond those indicated in the operating sections of this specification is not implied. Stresses greater than those listed may cause permanent damage to the device. Exposure to absolute maximum rating conditions for extended periods may affect reliability.

3. This device contains circuitry to protect the inputs against damage caused by high static voltages or electrical fields. However, it is advised that normal precautions be taken to avoid application of any voltage higher than the absolute maximum rated voltages to this high impedance circuit. Unused inputs should be tied to an appropriate logic level such as **VCC** or **GND**.

4. The input current applies to any input or output pin and applies when the voltage on the pin is between **GND** and **VCC**.

Table 10.2: Operating conditions

SYMBOL	PARAMETER	MIN	MAX	UNITS	NOTE
VCC	DC supply voltage	4.75	5.25	V	1
VI, VO	Input or output voltage	0	VCC	V	1,2
CL	Load capacitance on any pin		50	pF	
TA	Operating temperature range	0	70	°C	3

Notes

1. All voltages are with respect to **GND**.

2. Excursions beyond the supplies are permitted but not recommended; see DC characteristics.

3. Air flow rate 400 linear ft/min transverse air flow.

10 Electrical specifications

Table 10.3: DC characteristics

SYMBOL	PARAMETER		MIN	MAX	UNITS	NOTE
VIH	High level input voltage		2.0	VCC+0.5	V	1,2
VIL	Low level input voltage		-0.5	0.8	V	1,2
II	Input current	@ GND<VI<VCC		±10	µA	1,2
VOH	Output high voltage	@ IOH=2mA	VCC-1		V	1,2
VOL	Output low voltage	@ IOL=4mA		0.4	V	1,2
IOS	Output short circuit current	@ GND<VO<VCC		50	mA	1,2,4
				75	mA	1,2,5
IOZ	Tristate output current	@ GND<VO<VCC		±10	µA	1,2
PD	Power dissipation			700	mW	2,3
CIN	Input capacitance	@ f=1MHz		7	pF	
COZ	Output capacitance	@ f=1MHz		10	pF	

Notes

1 All voltages are with respect to **GND**.

2 Parameters measured at 4.75V<**VCC**<5.25V and 0°C<**TA**<70°C. Input clock frequency = 5MHz.

3 Power dissipation varies with output loading and program execution.

4 Current sourced from non-link outputs.

5 Current sourced from link outputs.

10.2 Equivalent circuits

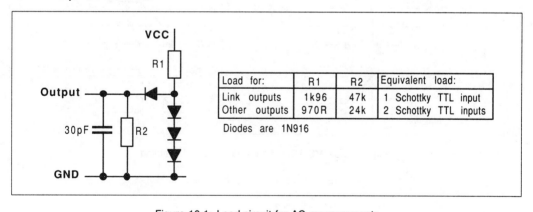

Figure 10.1: Load circuit for AC measurements

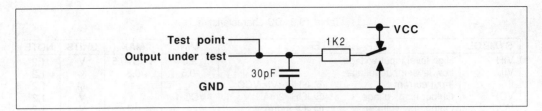

Figure 10.2: Tristate load circuit for AC measurements

10.3 AC timing characteristics

Table 10.4: Input, output edges

SYMBOL	PARAMETER	MIN	MAX	UNITS	NOTE
TDr	Input rising edges	2	20	ns	1,2
TDf	Input falling edges	2	20	ns	1,2
TQr	Output rising edges		25	ns	1
TQf	Output falling edges		15	ns	1

Notes

1 Non-link pins; see section on links.

2 All inputs except **ClockIn**; see section on **ClockIn**.

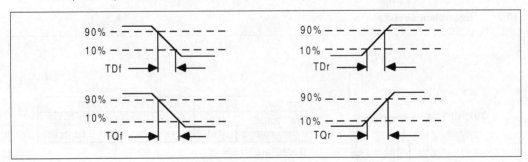

Figure 10.3: IMS T212 input and output edge timing

10 Electrical specifications

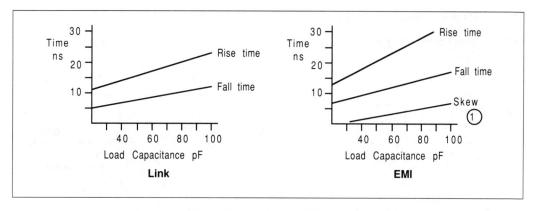

Figure 10.4: Typical rise/fall times

Notes

1. Skew is measured between **notMemCE** with a standard load (2 Schottky TTL inputs and 30pF) and **notMemCE** with a load of 2 Schottky TTL inputs and varying capacitance.

10.4 Power rating

Internal power dissipation P_{INT} of transputer and peripheral chips depends on **VCC**, as shown in figure 10.5. P_{INT} is substantially independent of temperature.

Total power dissipation P_D of the chip is

$$P_D = P_{INT} + P_{IO}$$

where P_{IO} is the power dissipation in the input and output pins; this is application dependent.

Internal working temperature T_J of the chip is

$$T_J = T_A + \theta J_A * P_D$$

where T_A is the external ambient temperature in °C and θJ_A is the junction-to-ambient thermal resistance in °C/W. θJ_A for each package is given in the Packaging Specifications section.

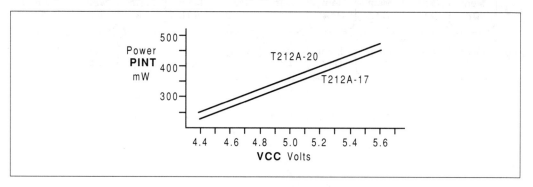

Figure 10.5: IMS T212 internal power dissipation vs VCC

11 Package specifications

11.1 68 pin grid array package

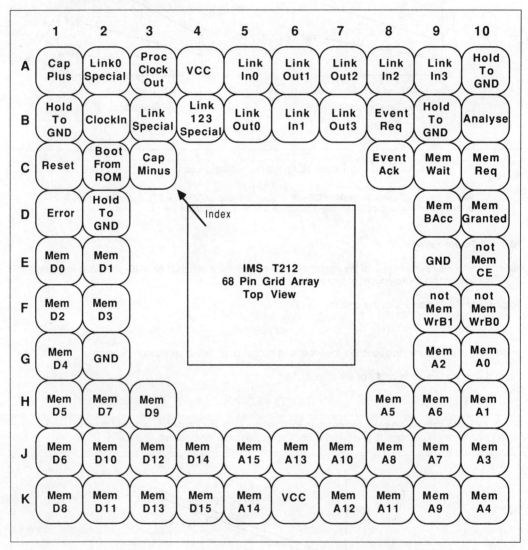

Figure 11.1: IMS T212 68 pin grid array package pinout

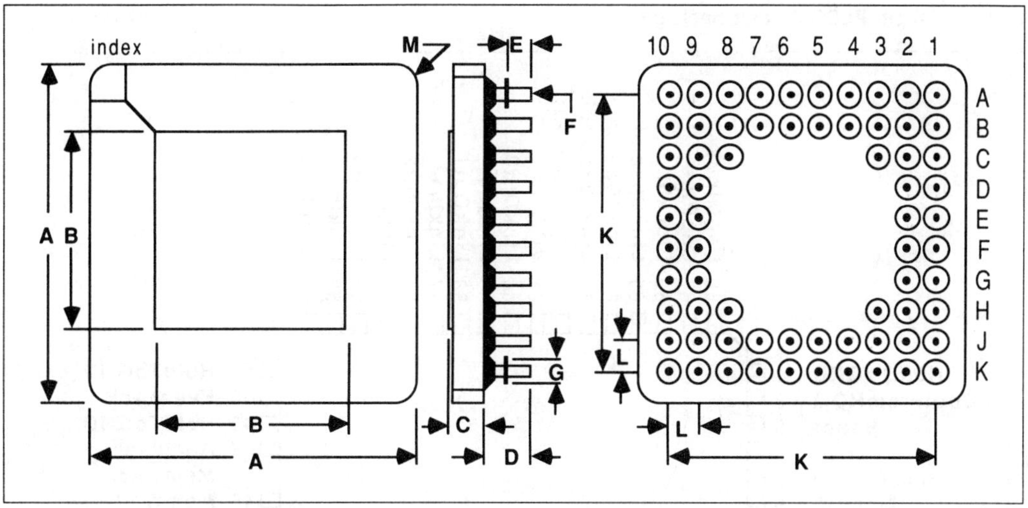

11 Package specifications

Figure 11.2: 68 pin grid array package dimensions

Table 11.1: 68 pin grid array package dimensions

	Millimetres		Inches		
DIM	NOM	TOL	NOM	TOL	Notes
A	26.924	±0.254	1.060	±0.010	
B	17.019	±0.127	0.670	±0.008	
C	2.466	±0.279	0.097	±0.011	
D	4.572	±0.127	0.180	±0.005	
E	3.302	±0.127	0.130	±0.005	
F	0.457	±0.051	0.018	±0.002	Pin diameter
G	1.270	±0.127	0.050	±0.005	Flange diameter
K	22.860	±0.127	0.900	±0.005	
L	2.540	±0.127	0.100	±0.005	
M	0.508		0.020		Chamfer

Package weight is approximately 6.8 grams

Table 11.2: 68 pin grid array package junction to ambient thermal resistance

SYMBOL	PARAMETER	MIN	NOM	MAX	UNITS	NOTE
θ_{JA}	At 400 linear ft/min transverse air flow			35	°C/W	

11.2 68 pin PLCC J-bend package

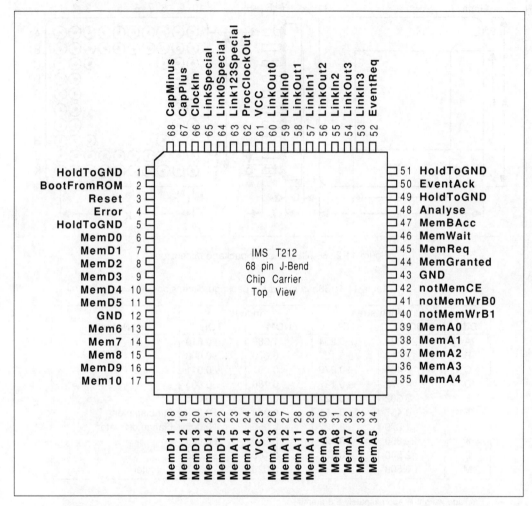

Figure 11.3: IMS T212 68 pin PLCC J-bend package pinout

11 Package specifications

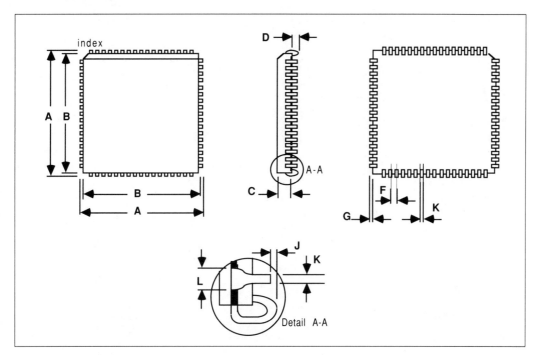

Figure 11.4: 68 pin PLCC J-bend package dimensions

Table 11.3: 68 pin PLCC J-bend package dimensions

DIM	Millimetres		Inches		Notes
	NOM	TOL	NOM	TOL	
A	25.146	±0.127	0.990	±0.005	
B	24.232	±0.127	0.954	±0.005	
C	3.810	±0.127	0.150	±0.005	
D	0.508	±0.127	0.020	±0.005	
F	1.270	±0.127	0.050	±0.005	
G	0.457	±0.127	0.018	±0.005	
J	0.000	±0.051	0.000	±0.002	
K	0.457	±0.127	0.018	±0.005	
L	0.762	±0.127	0.030	±0.005	

Package weight is approximately 5.0 grams

Table 11.4: 68 pin PLCC J-bend package junction to ambient thermal resistance

SYMBOL	PARAMETER	MIN	NOM	MAX	UNITS	NOTE
θJA	At 400 linear ft/min transverse air flow		35		°C/W	

Chapter 6

IMS M212 preview

1 Introduction

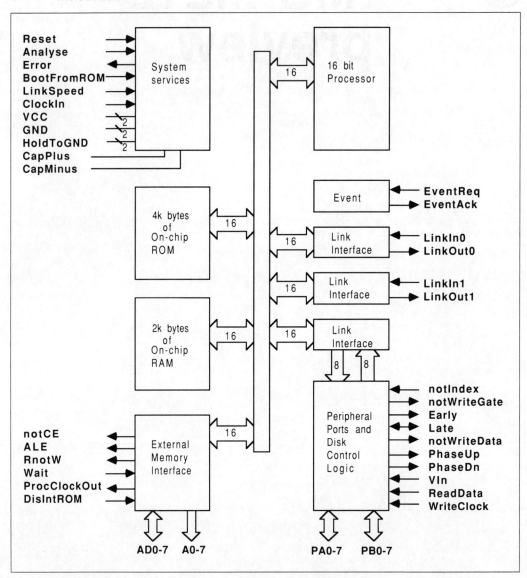

Figure 1.1: IMS M212 block diagram

1   Introduction

The IMS M212 peripheral processor is an intelligent peripheral controller of the INMOS transputer family, configured for connection to soft sectored winchester and floppy disk drives. It satisfies the demand for increasing intelligence in peripheral controllers and maintains a high degree of flexibility, allowing designers to modify the controller function without altering the hardware.

The disk control function has been designed to provide easy connection, with minimal external hardware, to a standard winchester and/or floppy disk interface. Two byte-wide programmable bidirectional ports are provided to control and monitor disk functions such as head position, drive selection and disk status. A dedicated port is provided for serial data interfaces and critical timing signals.

The IMS M212 is programmed as a normal transputer, permitting extremely powerful peripheral control facilities to be built into the device and thus reducing the load on the traditional central processor of a computer. Full details are given in the IMS M212 Disk Processor Product Data manual.

1.1   IMS M212 peripheral processor

1.1.1   Central processor

At the heart of the IMS M212 is the processor from the IMS T212 transputer. Its design achieves compact programs, efficient high level language implementation and provides direct support for the occam model of concurrency. The processor shares its time between any number of concurrent processes. A process waiting for communication or a timer does not consume any processor time. Two levels of process priority enable fast interrupt response to be achieved.

The IMS M212 has been designed so that the on-chip processor performs as many functions as possible, providing flexible operation and minimising on-chip disk-specific hardware.

1.1.2   Peripheral interface

The two 8 bit data ports **PA0-7** and **PB0-7** are controlled by the processor via a pair of channels. This allows the programmer to modify the function of these ports in order to implement a wide variety of applications.

The peripheral interface includes data output registers and TTL compatible input ports, as well as facilities for defining the direction of the pins on a bit-selectable basis. The interface contains logic to detect a change of state on the input pins and to store this change for interrogation by the program.

In addition to this, the external memory interface can support memory mapped peripherals on its byte-wide data bus. An event pin is also provided, so that peripherals can request attention.

1.1.3   Disk controller

The disk interface provides a simple interconnection to ST506/ST412 and SA400/SA450 compatible disk drives via ten dedicated disk control lines and the two general purpose 8 bit bidirectional data ports **PA0-7** and **PB0-7**. Although the on-chip disk control hardware handles much of the specialised data conversion, as many disk operations as possible are controlled by the processor, using sequences of control and data information.

The processor can program and interrogate all the registers controlling the disk functions and data ports, and thereby control the external interface lines. As a result of this versatility, the IMS M212 can also be used in applications other than disk control ones.

A versatile hardware 32 bit Error Correcting Codes (ECC) and 16 bit Cyclic Redundancy Codes (CRC) generator is included to check data integrity. ECC's allow certain classes of errors to be corrected as well as detected, whilst CRC's only allow detection.

When writing data to the disk the hardware serialises the data and encodes it into a Frequency Modulated (FM) or Modified Frequency Modulated (MFM) data stream. Any necessary precompensation is performed internally before outputting the data together with the necessary control signals. Any necessary modification of the data, for instance writing the Address Marks (AM) or inserting the CRC/ECC bytes, is automatically performed by the hardware.

When reading data from disk the raw read data is input and the function known as data separation is performed internally. The hardware examines the data stream for an Address Mark to achieve byte synchronisation and then searches for the desired sector information. When the required data is located it is decoded and a serial to parallel conversion is performed before the data is transferred to the processor.

1.1.4 Links

The IMS M212 uses a DMA block transfer mechanism to transfer messages between memory and another transputer product via the INMOS links. The link interfaces and the processor all operate concurrently, allowing processing to continue while data is being transferred on all of the links.

The host interface of the IMS M212 is via two INMOS standard links, providing simple connection to any transputer based system or, via a link adaptor, to a conventional microprocessor system. Link speeds of 10 Mbits/sec and 20 Mbits/sec are available, making the device compatible with all other INMOS transputer products.

The on-chip disk control logic is controlled by the processor, using simple command sequences, via two channels which appear to the processor as a normal pair of hardware channels.

1.1.5 Memory system

The 2 Kbytes of on-chip static RAM can be used for program or data storage, as a sector buffer or to store parameter and format information. It can be extended off chip, via the external memory interface, to provide a total of 64 Kbytes. Internal and external memory appear as a single contiguous address space.

Software contained in 4 Kbytes of internal ROM enables the IMS M212 to be used as a stand alone disk processor. The ROM can be disabled to free the address space for external memory.

1.1.6 Error handling

High level language execution is made secure with array bounds checking, arithmetic overflow detection etc. A flag is set when an error is detected, and the error can be handled internally by software or externally by sensing the error pin. System state is preserved for subsequent analysis.

2   Operation

The IMS M212 can be used in two modes: Mode 1, which uses the software in the internal ROM, and Mode 2, which relies upon custom designed software.

2.1   Mode 1

Mode 1 operation uses code in the on-chip ROM to control the disk controller hardware, and little knowledge of the hardware is required to implement winchester and floppy disk drivers. The programming interface to all drive types is identical, and there is sufficient flexibility to allow a wide variety of formats and drive types to be used.

Both ST506/412 compatible winchester and SA400/450 compatible floppy drives are supported in standard double density formats; this includes common 5.25 and 3.5 inch drives. Up to 4096 cylinders are allowed. Floppy drives can have up to 8 heads and winchesters up to 16 heads. There can be between 1 and 256 sectors per track, with sector sizes of 128 to 16384 bytes in powers of 2. Drives with or without 'seek complete' and 'ready' lines are supported, and step rates can be from $64\mu s$ to 16ms. A range of non-standard formats can also be set up for user-specific requirements.

As with transputers, the IMS M212 can be bootstrapped from ROM or via a link. In addition, the Mode 1 monitor process also provides a facility whereby the disk processor can bootstrap itself with code read from a disk; this code runs instead of the Mode 1 process. Another option sends a standard bootstrap message, read from a disk, out of link 0; the Mode 1 process then continues as normal. It is also possible in Mode 1 to send a command, at any time, to bootstrap from code in the sector buffer.

General workspace for Mode 1 is contained in on-chip RAM, which also provides 1280 bytes of sector buffer. Contiguous external RAM immediately past the internal RAM will automatically be used to extend the size of the sector buffer. As many sectors as will fit into the sector buffer can be stored in it at the same time.

In Mode 1 a separate data area, in on-chip RAM, contains all the required control information (parameters) for each of the four possible drives. Parameters may be read from or written to via the links, and contain such information as the capacity of the disk, current position of the heads, desired sector for reading or writing, drive type, timing details etc.

Command and data bytes are accepted down either of the IMS M212 links; an interlock system prevents conflict between commands received on both links simultaneously. Any results are returned on the link which received the command. Available commands are

EndOfSequence	Initialise	ReadParameter	WriteParameter
ReadBuffer	WriteBuffer	ReadSector	WriteSector
Restore	Seek	SelectHead	SelectDrive
PollDrives	FormatTrack	Boot	

Disk access commands implicitly select the drive, perform a seek and select the head. If an ECC or CRC error is found when reading a sector, a programmable number of automatic retries are performed and a subsequent correction attempted if possible. Mode 1 supports two of the four IMS M212 ECC/CRC modes - ECC and CRC. Either CRC or ECC can be specified in either of the ID or Data fields, making it possible to have floppies with correctable Data fields.

All appropriate parameters are checked to ensure that, for example, an attempt is not made to access a non-existent sector, relieving the host processor of such checking. Another feature which reduces the load on the host processor is the logical sector mode, in which all the sectors are specified as a single linear address space rather than physical cylinder/head/sector.

The logical address can also be auto-incremented if desired, as can the sector buffer. This allows a number of consecutive sectors to be read from or written to the disk with little overhead. As a *sticky status* checking technique is used, the status only has to be checked once at the end of a stream of commands; if an error occurred then reading and writing is inhibited, so that the logical address can be inspected to find where the error occurred.

2.2 Mode 2

In Mode 2 operation the internal ROM is bypassed, allowing the device to utilise user-defined software. This software can be held in external ROM, bootstrapped from a floppy or winchester disk, or loaded from the host processor via a link into internal or external RAM.

In this mode the user services the disk control hardware via a pair of on-chip high bandwidth channels. Using these channels the processor has access to the 49 registers which control the operation of the disk controller. Sequences of control codes and data bytes are sent by the processor to the disk controller logic via one of the hardware channels and data returned to the IMS M212 processor via the other. Each control code is a single byte, and may be followed by one or more data bytes.

In Mode 2 the designer can define new commands which are more complex than otherwise available. Examples include a *Format Disk* command as an extension to the *Format Track*; an application-specific directory structure; a software interface to optimise a particular file structure. Mode 2 also allows the user to optimise data transfer; thus, data could be read from a disk with no interleave, or data transfers could be re-ordered to minimise head movement. Disk searches can be arranged such that data transfer back to the host is minimised, as data comparisons can be performed by the on-chip processor.

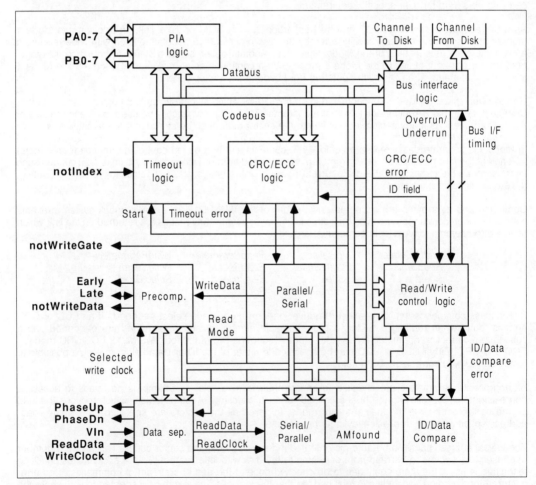

Figure 2.1: Disk controller interface

3 Applications

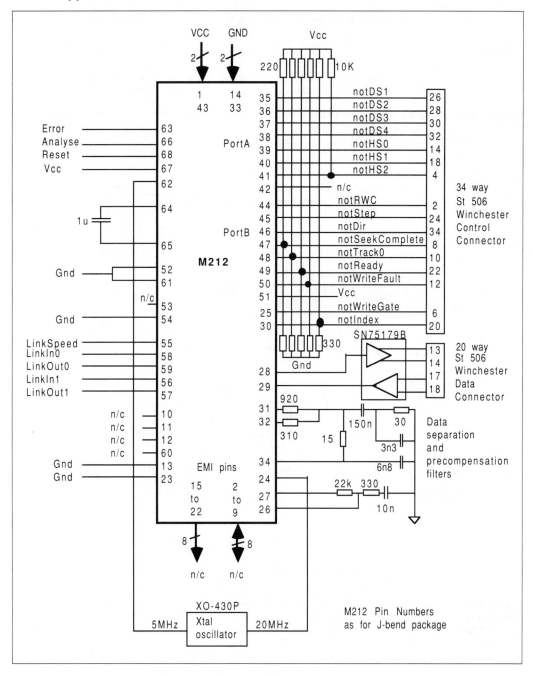

Figure 3.1: Winchester disk controller

The IMS M212 can interface to a floppy or winchester disk with very little external circuitry when used in Mode 1 or if a program is bootstrapped from a link. A typical arrangement is shown in figure 3.1. Note the absence of any control port buffers; this is possible provided the drive characteristics are not infringed.

Additional external memory can easily be added to the IMS M212. In both Modes 1 and 2, external RAM can be added for extra sector storage, whilst in Mode 2 extra RAM or ROM can be provided for program storage.

With the addition of control buffers and suitable clocks, a single IMS M212 can interface to both floppy and winchester drives. Link adaptors provide a means of interfacing to conventional microprocessors.

The IMS B005 evaluation board is an example of an application with control for both types of drive. The board also has a fully populated memory interface.

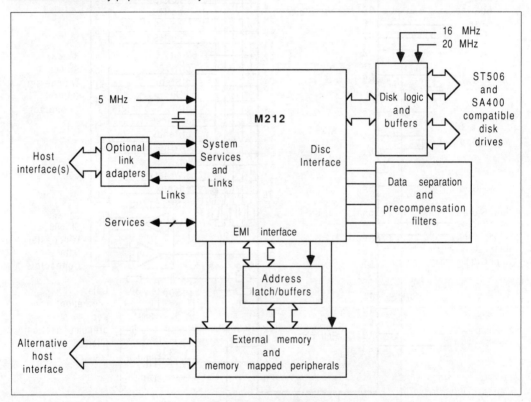

Figure 3.2: Enhanced disk controller interface

The IMS M212 can interface with both floppy and winchester disk drives, and the data rate to and from the disk can be selected by software. As a result the device is suitable for interfacing to the new generation of floppy disk drives which use vertical recording. These drives have an increased data rate of 1 Mbit/sec, and quadruple the capacity of existing floppy disk drives to 4 Mbytes. A single IMS M212 can be used to control a mixture of standard floppy drives, winchester drives and the new high speed high capacity drives. This eases compatibility and portability problems, and provides a simple upgrade path from standard floppies to high capacity floppies to winchesters.

3 Applications

The IMS M212 provides a very simple and compact disk controller solution, making it very easy to replace a single large disk drive with an array of IMS M212's, each controlling a single smaller disk drive. This has several advantages: cheaper drives can be used; overall available disk bandwidth is increased; local processing is provided by a high performance processor at each disk node; fault tolerant operation. The latter can be achieved by holding duplicated data on several drives. This prevents the whole system from stopping, as would be the case if the single large drive failed.

These advantages are particularly applicable when transputers are connected in arrays to provide high performance concurrent systems (figure 3.3). The IMS M212's can be directly connected to the array via INMOS links and the spare link used to communicate with the adjacent IMS M212 to provide the fault tolerant operation.

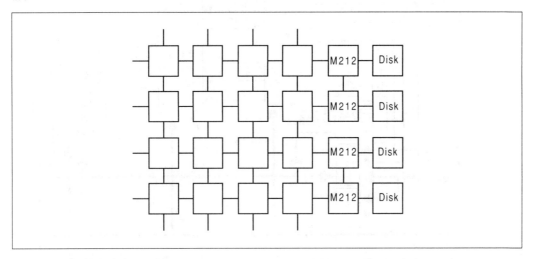

Figure 3.3: Transputer network with disk processors

A high performance processor allows many operations to be performed locally to the disk. This not only frees the host processor for other work but also removes the need for large amounts of data to be needlessly transferred to the host. Operations which can be performed by the IMS M212 include: file management with directory management and pre-reading; data manipulation such as compression/de-compression and encryption/de-cryption; data search such as database key searching; performance optimisation such as head scheduling and cacheing.

The IMS M212 external memory interface can be used to connect to memory mapped peripherals. One application of this is interfacing to a SCSI bus controller, permitting direct connection to the SCSI bus in a low part count system. The processor is used to control the SCSI bus controller and implements the required command interface, as well as controlling the disk or other peripheral.

This arrangement allows floppy and winchester disks to be simply connected to a SCSI bus. Because the command interface is controlled by a process running in the IMS M212, any future command upgrades can easily be incorporated.

The design can be used both as a target and an initiator interface, again controlled by the process running in the IMS M212. It provides a means of implementing a link to SCSI interface, as well as a SCSI controlled disk.

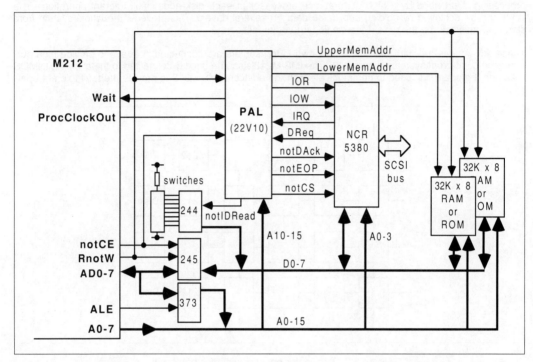

Figure 3.4: SCSI interface

4 Package specifications

4.1 68 pin grid array package

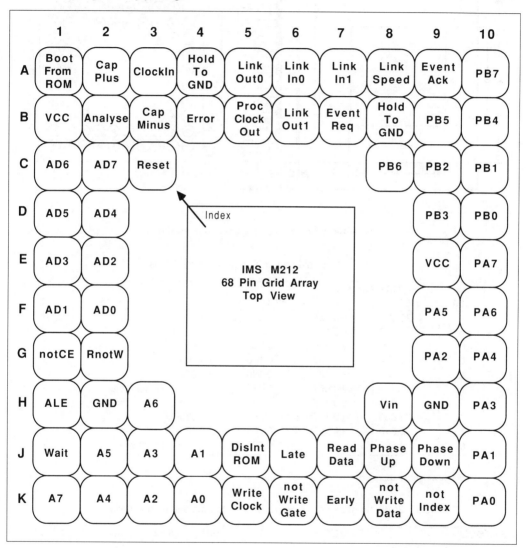

Figure 4.1: IMS M212 68 pin grid array package pinout

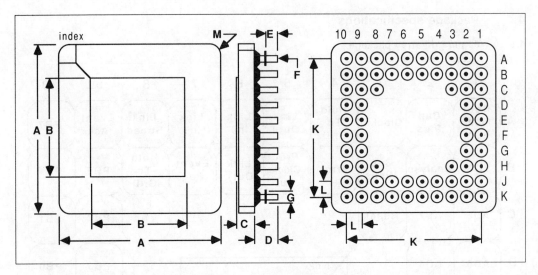

Figure 4.2: 68 pin grid array package dimensions

Table 4.1: 68 pin grid array package dimensions

DIM	Millimetres		Inches		Notes
	NOM	TOL	NOM	TOL	
A	26.924	±0.254	1.060	±0.010	
B	17.019	±0.127	0.670	±0.008	
C	2.466	±0.279	0.097	±0.011	
D	4.572	±0.127	0.180	±0.005	
E	3.302	±0.127	0.130	±0.005	
F	0.457	±0.051	0.018	±0.002	Pin diameter
G	1.270	±0.127	0.050	±0.005	Flange diameter
K	22.860	±0.127	0.900	±0.005	
L	2.540	±0.127	0.100	±0.005	
M	0.508		0.020		Chamfer

Package weight is approximately 6.8 grams

Table 4.2: 68 pin grid array package junction to ambient thermal resistance

SYMBOL	PARAMETER	MIN	NOM	MAX	UNITS	NOTE
θJA	At 400 linear ft/min transverse air flow			35	°C/W	

4 Package specifications

4.2 68 pin PLCC J-bend package

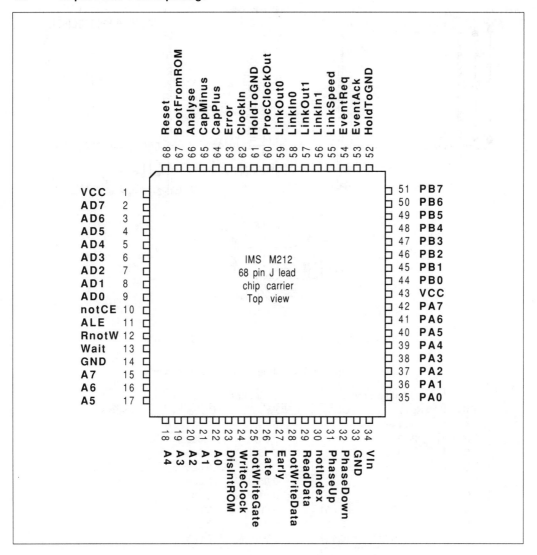

Figure 4.3: IMS M212 68 pin PLCC J-bend package pinout

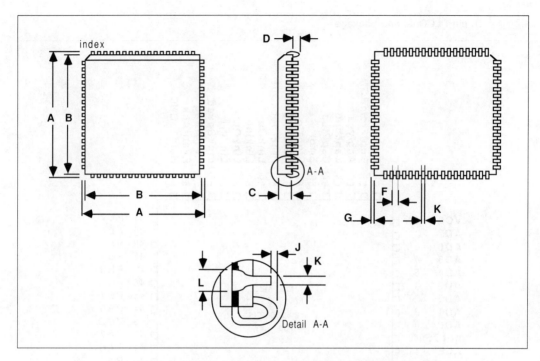

Figure 4.4: 68 pin PLCC J-bend package dimensions

Table 4.3: 68 pin PLCC J-bend package dimensions

DIM	Millimetres		Inches		Notes
	NOM	TOL	NOM	TOL	
A	25.146	±0.127	0.990	±0.005	
B	24.232	±0.127	0.954	±0.005	
C	3.810	±0.127	0.150	±0.005	
D	0.508	±0.127	0.020	±0.005	
F	1.270	±0.127	0.050	±0.005	
G	0.457	±0.127	0.018	±0.005	
J	0.000	±0.051	0.000	±0.002	
K	0.457	±0.127	0.018	±0.005	
L	0.762	±0.127	0.030	±0.005	

Package weight is approximately 5.0 grams

Table 4.4: 68 pin PLCC J-bend package junction to ambient thermal resistance

SYMBOL	PARAMETER	MIN	NOM	MAX	UNITS	NOTE
θJA	At 400 linear ft/min transverse air flow		35		°C/W	

Chapter 7

IMS C004 engineering data

1 Introduction

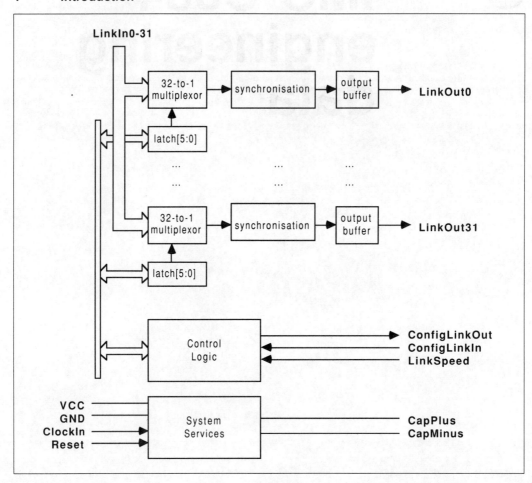

Figure 1.1: IMS C004 block diagram

1 Introduction

The INMOS communication link is a high speed system interconnect which provides full duplex communication between members of the INMOS transputer family, according to the INMOS serial link protocol. The IMS C004, a member of this family, is a transparent programmable link switch designed to provide a full crossbar switch between 32 link inputs and 32 link outputs.

The IMS C004 will switch links running at either the standard speed of 10 Mbits/sec or at the higher speed of 20 Mbit/sec. It introduces, on average, only a 1.75 bit time delay on the signal. Link switches can be cascaded to any depth without loss of signal integrity and can be used to construct reconfigurable networks of arbitrary size. The switch is programmed via a separate serial link called the *configuration link*.

All INMOS products which use communication links, regardless of device type, support a standard communications frequency of 10 Mbits/sec; most products also support 20 Mbits/sec. Products of different type or performance can, therefore, be interconnected directly and future systems will be able to communicate directly with those of today.

2 Pin designations

Table 2.1: IMS C004 services services

Pin	In/Out	Function
VCC, GND		Power supply and return
CapPlus, CapMinus		External capacitor for internal clock power supply
ClockIn	in	Input clock
Reset	in	System reset
DoNotWire		Must not be wired

Table 2.2: IMS C004 configuration

Pin	In/Out	Function
ConfigLinkIn	in	INMOS configuration link input
ConfigLinkOut	out	INMOS configuration link output

Table 2.3: IMS C004 link

Pin	In/Out	Function
LinkIn0-31	in	INMOS link inputs to the switch
LinkOut0-31	out	INMOS link outputs from the switch
LinkSpeed	in	Link speed selection

Signal names are prefixed by **not** if they are active low, otherwise they are active high.
Pinout details for various packages are given on page 245.

3 System services

System services include all the necessary logic to start up and maintain the IMS C004.

3.1 Power

Power is supplied to the device via the **VCC** and **GND** pins. Several of each are provided to minimise inductance within the package. All supply pins must be connected. The supply must be decoupled close to the chip by at least one 100nF low inductance (e.g. ceramic) capacitor between **VCC** and **GND**. Four layer boards are recommended; if two layer boards are used, extra care should be taken in decoupling.

Input voltages must not exceed specification with respect to **VCC** and **GND**, even during power-up and power-down ramping, otherwise *latchup* can occur. CMOS devices can be permanently damaged by excessive periods of latchup.

3.2 CapPlus, CapMinus

The internally derived power supply for internal clocks requires an external low leakage, low inductance 1 μF capacitor to be connected between **CapPlus** and **CapMinus**. A ceramic capacitor is preferred, with an impedance less than 3 ohms between 100 KHz and 10 MHz. If a polarised capacitor is used the negative terminal should be connected to **CapMinus**. Total PCB track length should be less than 50mm. The connections must not touch power supplies or other noise sources.

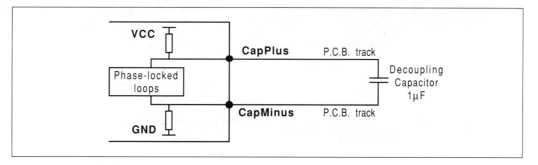

Figure 3.1: Recommended PLL decoupling

3.3 ClockIn

Transputer family components use a standard clock frequency, supplied by the user on the **ClockIn** input. The nominal frequency of this clock for all transputer family components is 5MHz, regardless of device type, transputer word length or processor cycle time. High frequency internal clocks are derived from **ClockIn**, simplifying system design and avoiding problems of distributing high speed clocks externally.

A number of transputer family devices may be connected to a common clock, or may have individual clocks providing each one meets the specified stability criteria. In a multi-clock system the relative phasing of **ClockIn** clocks is not important, due to the asynchronous nature of the links. Mark/space ratio is unimportant provided the specified limits of **ClockIn** pulse widths are met.

Oscillator stability is important. **ClockIn** must be derived from a crystal oscillator; RC oscillators are not sufficiently stable. **ClockIn** must not be distributed through a long chain of buffers. Clock edges must be monotonic and remain within the specified voltage and time limits.

Table 3.1: Input clock

SYMBOL	PARAMETER	MIN	NOM	MAX	UNITS	NOTE
TDCLDCH	ClockIn pulse width low	40			ns	
TDCHDCL	ClockIn pulse width high	40			ns	
TDCLDCL	ClockIn period		200		ns	1,3
TDCerror	ClockIn timing error			±0.5	ns	2
TDC1DC2	Difference in ClockIn for 2 linked devices			400	ppm	3
TDCr	ClockIn rise time			10	ns	4
TDCf	ClockIn fall time			8	ns	4

Notes

1 Measured between corresponding points on consecutive falling edges.

2 Variation of individual falling edges from their nominal times.

3 This value allows the use of 200ppm crystal oscillators for two devices connected together by a link.

4 Clock transitions must be monotonic within the range **VIH** to **VIL** (page 242).

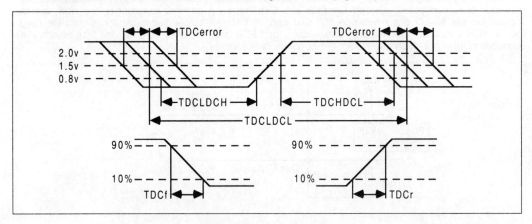

Figure 3.2: ClockIn timing

3 System services

3.4 Reset

The **Reset** pin can go high with **VCC**, but must at no time exceed the maximum specified voltage for **VIH**. After **VCC** is valid **ClockIn** should be running for a minimum period **TDCVRL** before the end of **Reset**.

Reset initialises the IMS C004 to a state where all link outputs from the switch are disconnected and held low; the control link is then ready to receive a configuration message.

Table 3.2: Reset

SYMBOL	PARAMETER	MIN	NOM	MAX	UNITS	NOTE
TPVRH	Power valid before Reset	10			ms	
TRHRL	Reset pulse width high	8			ClockIn	1
TDCVRL	ClockIn running before Reset end	10			ms	2

Notes

1 Full periods of **ClockIn TDCLDCL** required.

2 At power-on reset.

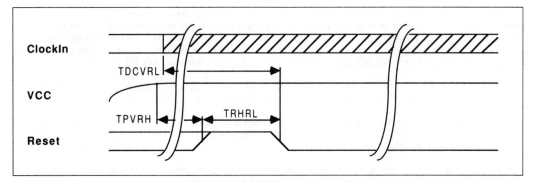

Figure 3.3: Reset timing

4   Links

INMOS bi-directional serial links provide synchronized communication between INMOS products and with the outside world. Each link comprises an input channel and output channel. A link between two devices is implemented by connecting a link interface on one device to a link interface on the other device. Every byte of data sent on a link is acknowledged on the input of the same link, thus each signal line carries both data and control information.

A receiver can transmit an acknowledge as soon as it starts to receive a data byte. In this way the transmission of an acknowledge can be overlapped with receipt of a data byte to provide continuous transmission of data. This technique is fully compatible with all other INMOS transputer family links.

The quiescent state of a link output is low. Each data byte is transmitted as a high start bit followed by a one bit followed by eight data bits followed by a low stop bit. The least significant bit of data is transmitted first. After transmitting a data byte the sender waits for the acknowledge, which consists of a high start bit followed by a zero bit. The acknowledge signifies that the receiving link is able to receive another byte.

Links are not synchronised with **ClockIn** and are insensitive to its phase. Thus links from independently clocked systems may communicate, providing only that the clocks are nominally identical and within specification.

Links are TTL compatible and intended to be used in electrically quiet environments, between devices on a single printed circuit board or between two boards via a backplane. Direct connection may be made between devices separated by a distance of less than 300 millimetres. For longer distances a matched 100 Ohm transmission line should be used with series matching resistors **RM**. When this is done the line delay should be less than 0.4 bit time to ensure that the reflection returns before the next data bit is sent.

Buffers may be used for very long transmissions. If so, their overall propagation delay should be stable within the skew tolerance of the link, although the absolute value of the delay is immaterial.

The IMS C004 links support the standard INMOS communication speed of 10 Mbits per second. In addition they can be used at 20 Mbits per second. When the **LinkSpeed** pin is low, all links operate at the standard 10 Mbits/sec; when high they operate at 20 Mbits/sec.

A single IMS C004 inserted between two transputers which fully impelement overlapped acknowledges causes no reduction in data bandwidth, the delay through the switch being hidden by the overlapped acknowledge.

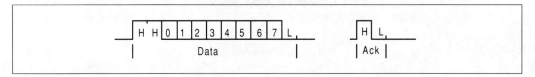

Figure 4.1: IMS C004 link data and acknowledge packets

4 Links

Table 4.1: Link

SYMBOL	PARAMETER		MIN	NOM	MAX	UNITS	NOTE
TJQr	LinkOut rise time				20	ns	
TJQf	LinkOut fall time				10	ns	
TJDr	LinkIn rise time				20	ns	
TJDf	LinkIn fall time				20	ns	
TJQJD	Buffered edge delay		0			ns	
TJBskew	Variation in TJQJD	20 Mbits/s			3	ns	1
		10 Mbits/s			10	ns	1
CLIZ	LinkIn capacitance	@ f=1MHz			7	pF	
CLL	LinkOut load capacitance				50	pF	
RM	Series resistor for 100Ω transmission line			56		ohms	

Notes

1 This is the variation in the total delay through buffers, transmission lines, differential receivers etc., caused by such things as short term variation in supply voltages and differences in delays for rising and falling edges.

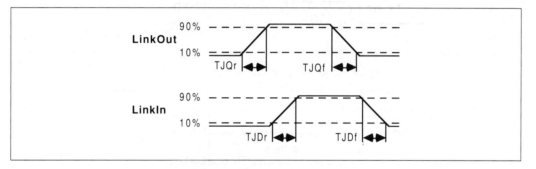

Figure 4.2: IMS C004 link timing

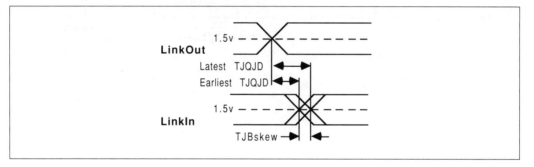

Figure 4.3: IMS C004 buffered link timing

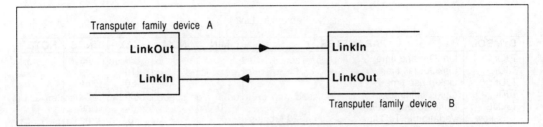

Figure 4.4: Links directly connected

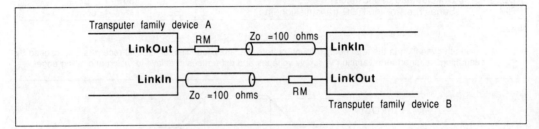

Figure 4.5: Links connected by transmission line

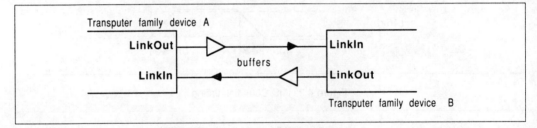

Figure 4.6: Links connected by buffers

5 Switch implementation

The IMS C004 is internally organised as a set of thirtytwo 32-to-1 multiplexors. Each multiplexor has associated with it a six bit latch, five bits of which select one input as the source of data for the corresponding output. The sixth bit is used to connect and disconnect the output. These latches can be read and written by messages sent on the configuration link via **ConfigLinkIn** and **ConfigLinkOut**.

The output of each multiplexor is synchronised with an internal high speed clock and regenerated at the output pad. This synchronisation introduces, on average, a 1.75 bit time delay on the signal. As the signal is not electrically degraded in passing through the switch, it is possible to form links through an arbitrary number of link switches.

Each input and output is identified by a number in the range 0 to 31. A configuration message consisting of one, two or three bytes is transmitted on the configuration link. The configuration messages sent to the switch on this link are shown in table 5.1. If an unspecified configuration message is used, the effect of it is undefined. A subset of table 5.1 is implemented on the IMS C004-A; see page 247.

Table 5.1: IMS C004 configuration messages

Configuration Message	Function
[0] [input] [output]	Connects **input** to **output**.
[1] [link1] [link2]	Connects **link1** to **link2** by connecting the input of **link1** to the output of **link2** and the input of **link2** to the output of **link1**.
[2] [output]	Enquires which input the **output** is connected to. The IMS C004 responds with the input. The most signifigant bit of this byte indicates whether the output is connected (bit set high) or disconnected (bit set low).
[3]	This command byte must be sent at the end of every configuration sequence which sets up a connection. The IMS C004 is then ready to accept data on the connected inputs.
[4]	Resets the switch. All outputs are disconnected and held low. This also happens when **Reset** is applied to the IMS C004.
[5] [output]	Output **output** is disconnected and held low.
[6] [link1] [link2]	Disconnects the output of **link1** and the output of **link2**.

6   Applications

6.1   Link switching

The IMS C004 provides full switching capabilities between 32 INMOS links. It can also be used as a component of a larger link switch. For example, three IMS C004's can be connected together to produce a 48 way switch, as shown in figure 6.1. This technique can be extended to the switch shown in figure 6.2.

A fully connected network of 32 INMOS transputers (one in which all four links are used on every transputer) can be completely configured using just four IMS C004's. Figure 6.5 shows the connected transputer network.

In these diagrams each link line shown represents a unidirectional link; i.e. one output to one input. Where a number is also given, that denotes the number of lines.

6.2   Multiple IMS C004 control

Many systems require a number of IMS C004's, each configured via its own configuration link. A simple method of implementing this uses a master IMS C004, as shown in figure 6.3. One of the transputer links is used to configure the master link switch, whilst another transputer link is multiplexed via the master to send configuration messages to any of the other 31 IMS C004 links.

6.3   Bidirectional exchange

Use of the IMS C004 is not restricted to computer configuration applications. The ability to change the switch setting dynamically enables it to be used as a general purpose message router. This may, of course, also find applications in computing with the emergence of the new generation of supercomputers, but a more widespread use may be found as a communication exchange.

In the application shown in figure 6.4, a message into the exchange must be preceded by a destination token *dest*. When this message is passed, the destination token is replaced with a source token so that the receiver knows where the message has come from. The **in.out** device in the diagram and the controller can be implemented easily with a transputer, and the link protocol for establishing communication with these devices can be interfaced with INMOS link adaptors. All messages from **rx[i]** are preceded by the destination output *dest*. On receipt of such a message the **in.out** device requests the controller to connect a bidirectional link path to *dest*. The controller determines what is currently connected to each end of the proposed link. When both ends are free it sets up the IMS C004 and informs both ends of the new link. Note that in this network two channels are placed on each IMS C004 link, one for each direction.

6.4   Bus systems

The IMS C004 can be used in conjunction with the INMOS IMS C011/C012 link adaptors to provide a flexible means of connecting conventional bus based microprocessor systems.

6 Applications

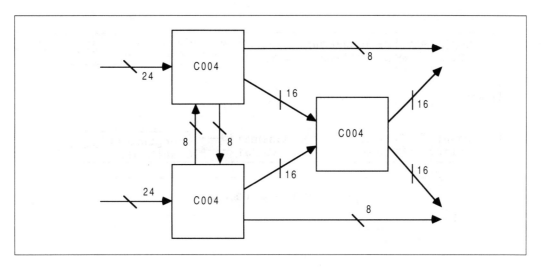

Figure 6.1: 48 way link switch

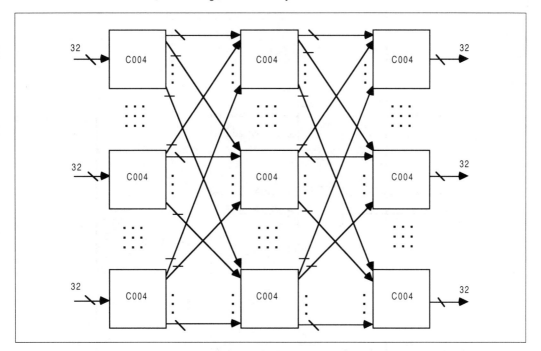

Figure 6.2: Generalised link switch

7 IMS C004 engineering data

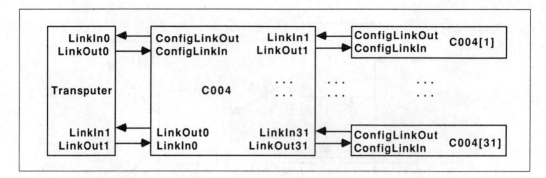

Figure 6.3: Multiple IMS C004 controller

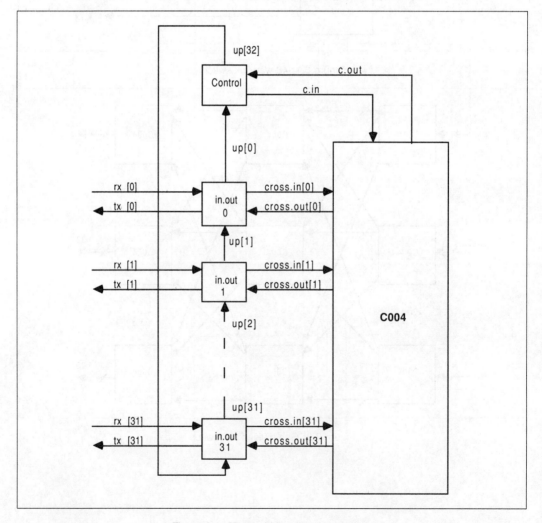

Figure 6.4: 32 way bidirectional exchange

6 Applications

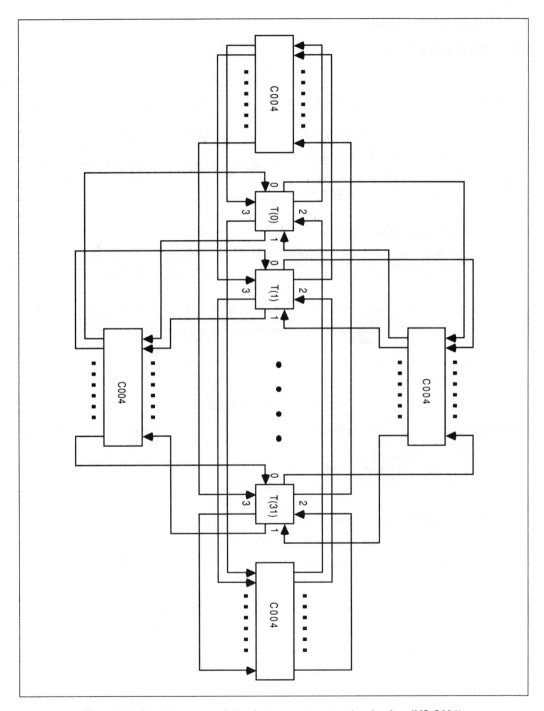

Figure 6.5: Complete connectivity of a transputer network using four IMS C004's

7 Electrical specifications

7.1 DC electrical characteristics

Table 7.1: Absolute maximum ratings

SYMBOL	PARAMETER	MIN	MAX	UNITS	NOTE
VCC	DC supply voltage	0	7.0	V	1,2,3
VI, VO	Voltage on input and output pins	-0.5	VCC+0.5	V	1,2,3
II	Input current		±25	mA	4
OSCT	Output short circuit time (one pin)		1	s	2
TS	Storage temperature	-65	150	°C	2
TA	Ambient temperature under bias	-55	125	°C	2
PDmax	Maximum allowable dissipation		2	W	

Notes

1. All voltages are with respect to **GND**.

2. This is a stress rating only and functional operation of the device at these or any other conditions beyond those indicated in the operating sections of this specification is not implied. Stresses greater than those listed may cause permanent damage to the device. Exposure to absolute maximum rating conditions for extended periods may affect reliability.

3. This device contains circuitry to protect the inputs against damage caused by high static voltages or electrical fields. However, it is advised that normal precautions be taken to avoid application of any voltage higher than the absolute maximum rated voltages to this high impedance circuit. Unused inputs should be tied to an appropriate logic level such as **VCC** or **GND**.

4. The input current applies to any input or output pin and applies when the voltage on the pin is between **GND** and **VCC**.

Table 7.2: Operating conditions

SYMBOL	PARAMETER	MIN	MAX	UNITS	NOTE
VCC	DC supply voltage	4.75	5.25	V	1
VI, VO	Input or output voltage	0	VCC	V	1,2
CL	Load capacitance on any pin		50	pF	
TA	Operating temperature range	0	70	°C	3

Notes

1. All voltages are with respect to **GND**.

2. Excursions beyond the supplies are permitted but not recommended; see DC characteristics.

3. Air flow rate 400 linear ft/min transverse air flow.

7 Electrical specifications

Table 7.3: DC characteristics

SYMBOL	PARAMETER		MIN	MAX	UNITS	NOTE
VIH	High level input voltage		2.0	VCC+0.5	V	1,2
VIL	Low level input voltage		-0.5	0.8	V	1,2
II	Input current	@ GND<VI<VCC		±10	µA	1,2
VOH	Output high voltage	@ IOH=2mA	VCC-1		V	1,2
VOL	Output low voltage	@ IOL=4mA		0.4	V	1,2
IOS	Output short circuit current	@ GND<VO<VCC		50	mA	1,2,4
				75	mA	1,2,5
PD	Power dissipation			1.5	W	2,3
CIN	Input capacitance	@ f=1MHz		7	pF	
COZ	Output capacitance	@ f=1MHz		10	pF	

Notes

1. All voltages are with respect to **GND**.
2. Parameters measured at 4.75V<**VCC**<5.25V and 0°C<**TA**<70°C. Input clock frequency = 5MHz.
3. Power dissipation varies with output loading and with the number of links active.
4. Current sourced from non-link outputs.
5. Current sourced from link outputs.

7.2 Equivalent circuits

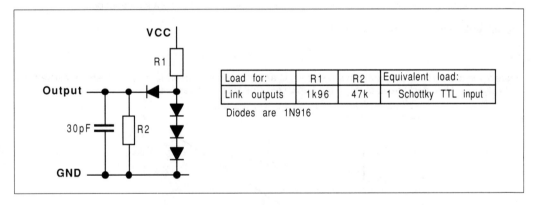

Figure 7.1: Load circuit for AC measurements

7.3 AC timing characteristics

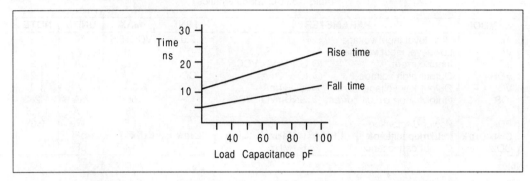

Figure 7.2: Typical link rise/fall times

7.4 Power rating

Internal power dissipation P_{INT} of transputer and peripheral chips depends on **VCC**, as shown in figure 7.3. P_{INT} is substantially independent of temperature.

Total power dissipation P_D of the chip is

$$P_D = P_{INT} + P_{IO}$$

where P_{IO} is the power dissipation in the input and output pins; this is application dependent.

Internal working temperature T_J of the chip is

$$T_J = T_A + \theta J_A * P_D$$

where T_A is the external ambient temperature in °C and θJ_A is the junction-to-ambient thermal resistance in °C/W. θJ_A for each package is given in the Packaging Specifications section.

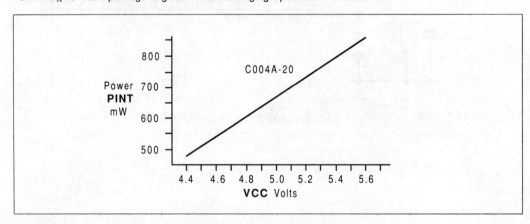

Figure 7.3: IMS C004 internal power dissipation vs VCC

8 Package specifications

8.1 84 pin grid array package

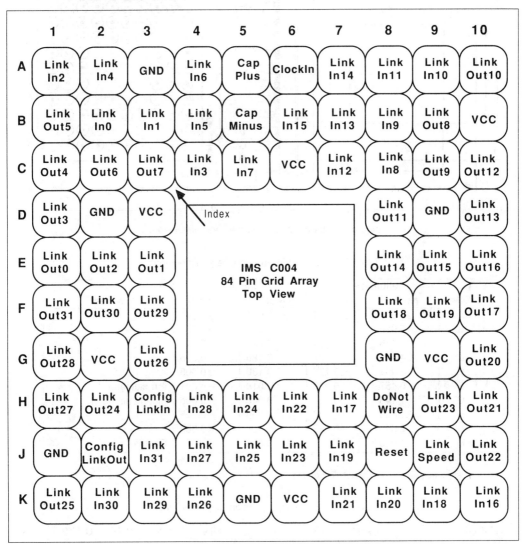

Figure 8.1: IMS C004 84 pin grid array package pinout

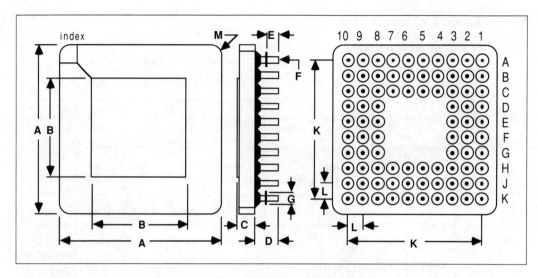

Figure 8.2: 84 pin grid array package dimensions

Table 8.1: 84 pin grid array package dimensions

DIM	Millimetres		Inches		Notes
	NOM	TOL	NOM	TOL	
A	26.924	±0.254	1.060	±0.010	
B	17.019	±0.127	0.670	±0.005	
C	2.456	±0.278	0.097	±0.011	
D	4.572	±0.127	0.180	±0.005	
E	3.302	±0.127	0.130	±0.005	
F	0.457	±0.025	0.018	±0.001	Pin diameter
G	1.143	±0.127	0.045	±0.005	Flange diameter
K	22.860	±0.127	0.900	±0.005	
L	2.540	±0.127	0.100	±0.005	
M	0.508		0.020		Chamfer

Package weight is approximately 7.2 grams

Table 8.2: 84 pin grid array package junction to ambient thermal resistance

SYMBOL	PARAMETER	MIN	NOM	MAX	UNITS	NOTE
θJA	At 400 linear ft/min transverse air flow			35	°C/W	

9 IMS C004-A

The IMS C004-A implements a subset of the configuration messages.

Table 9.1: IMS C004 configuration messages

Configuration Message	Function
[0] [input] [output]	Connects **input** to **output**.
[1] [link1] [link2]	Connects **link1** to **link2** by connecting the input of **link1** to the output of **link2** and the input of **link2** to the output of **link1**.
[2] [output]	Enquires which input the **output** is connected to. The IMS C004 responds with the input.
[3]	This command byte must be sent at the end of every configuration sequence which sets up a connection. The IMS C004 is then ready to accept data on the connected inputs.
[4]	Resets the switch. All outputs are disconnected and held low.

When **Reset** is applied to the IMS C004-A the outputs are not disconnected. After power is applied and before any configuration message is transmitted to the IMS C004-A, a software reset byte (control byte **[4]**) must be sent. This has the effect of disconnecting all the outputs.

inmos

Chapter 8

IMS C011 engineering data

1 Introduction

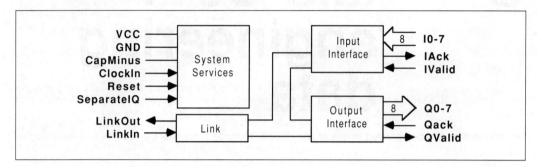

Figure 1.1: IMS C011 Mode 1 block diagram

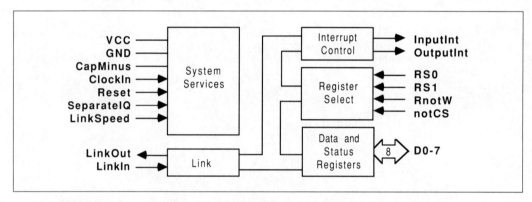

Figure 1.2: IMS C011 Mode 2 block diagram

1 Introduction

The INMOS communication link is a high speed system interconnect which provides full duplex communication between members of the INMOS transputer family, according to the INMOS serial link protocol. The IMS C011, a member of this family, provides for full duplex transputer link communication with standard microprocessor and sub-system architectures, by converting bi-directional serial link data into parallel data streams.

All INMOS products which use communication links, regardless of device type, support a standard communications frequency of 10 Mbits/sec; most products also support 20 Mbits/sec. Products of different type or performance can, therefore, be interconnected directly and future systems will be able to communicate directly with those of today. The IMS C011 link will run at either the standard speed of 10 Mbits/sec or at the higher speed of 20 Mbit/sec. Data reception is asynchronous, allowing communication to be independent of clock phase.

The link adaptor can be operated in one of two modes. In Mode 1 the IMS C011 converts between a link and two independent fully handshaken byte-wide interfaces, one input and one output. It can be used by a peripheral device to communicate with a transputer, an INMOS peripheral processor or another link adaptor, or it can provide programmable input and output pins for a transputer. Two IMS C011 devices in this mode can be connected back to back via the parallel ports and used as a frequency changer between different speed links.

In Mode 2 the IMS C011 provides an interface between an INMOS serial link and a microprocessor system bus. Status and data registers for both input and output ports can be accessed across the byte-wide bi-directional interface. Two interrupt outputs are provided, one to indicate input data available and one for output buffer empty.

2 Pin designations

Table 2.1: IMS C011 services and link

Pin	In/Out	Function
VCC, GND		Power supply and return
CapMinus		External capacitor for internal clock power supply
ClockIn	in	Input clock
Reset	in	System reset
SeparateIQ	in	Select mode and Mode 1 link speed
LinkIn	in	Serial data input channel
LinkOut	out	Serial data output channel

Table 2.2: IMS C011 Mode 1 parallel interface

Pin	In/Out	Function
I0-7	in	Parallel input bus
IValid	in	Data on **I0-7** is valid
IAck	out	Acknowledge **I0-7** data received by other link
Q0-7	out	Parallel output bus
QValid	out	Data on **Q0-7** is valid
QAck	in	Acknowledge from device: data **Q0-7** was read

Table 2.3: IMS C011 Mode 2 parallel interface

Pin	In/Out	Function
D0-7	in/out	Bi-directional data bus
notCS	in	Chip select
RS0-1	in	Register select
RnotW	in	Read/write control signal
InputInt	out	Interrupt on link receive buffer full
OutputInt	out	Interrupt on link transmit buffer empty
LinkSpeed	in	Select link speed as 10 or 20 Mbits/sec
HoldToGND		Must be connected to **GND**
DoNotWire		Must not be wired

Signal names are prefixed by **not** if they are active low, otherwise they are active high.
Pinout details for various packages are given on page 270.

3 System services

System services include all the necessary logic to start up and maintain the IMS C011.

3.1 Power

Power is supplied to the device via the **VCC** and **GND** pins. The supply must be decoupled close to the chip by at least one 100nF low inductance (e.g. ceramic) capacitor between **VCC** and **GND**. Four layer boards are recommended; if two layer boards are used, extra care should be taken in decoupling.

AC noise between **VCC** and **GND** must be kept below 200 mV peak to peak at all frequencies above 100 KHz. AC noise between **VCC** and the ground reference of load capacitances must be kept below 200 mV peak to peak at all frequencies above 30 MHz. Input voltages must not exceed specification with respect to **VCC** and **GND**, even during power-up and power-down ramping, otherwise *latchup* can occur. CMOS devices can be permanently damaged by excessive periods of latchup.

3.2 CapMinus

The internally derived power supply for internal clocks requires an external low leakage, low inductance 1 μF capacitor to be connected between **VCC** and **CapMinus**. A ceramic capacitor is preferred, with an impedance less than 3 ohms between 100 KHz and 10 MHz. If a polarised capacitor is used the negative terminal should be connected to **CapMinus**. Total PCB track length should be less than 50mm. The positive connection of the capacitor must be connected directly to **VCC**. Connections must not otherwise touch power supplies or other noise sources.

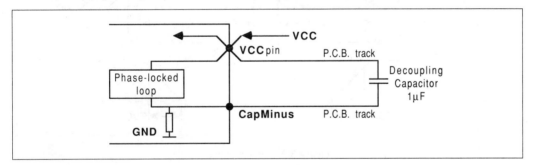

Figure 3.1: Recommended PLL decoupling

3.3 ClockIn

Transputer family components use a standard clock frequency, supplied by the user on the **ClockIn** input. The nominal frequency of this clock for all transputer family components is 5MHz, regardless of device type, transputer word length or processor cycle time. High frequency internal clocks are derived from **ClockIn**, simplifying system design and avoiding problems of distributing high speed clocks externally.

A number of transputer family devices may be connected to a common clock, or may have individual clocks providing each one meets the specified stability criteria. In a multi-clock system the relative phasing of **ClockIn** clocks is not important, due to the asynchronous nature of the links. Mark/space ratio is unimportant provided the specified limits of **ClockIn** pulse widths are met.

Oscillator stability is important. **ClockIn** must be derived from a crystal oscillator; RC oscillators are not sufficiently stable. **ClockIn** must not be distributed through a long chain of buffers. Clock edges must be monotonic and remain within the specified voltage and time limits.

Table 3.1: Input clock

SYMBOL	PARAMETER	MIN	NOM	MAX	UNITS	NOTE
TDCLDCH	ClockIn pulse width low	40			ns	
TDCHDCL	ClockIn pulse width high	40			ns	
TDCLDCL	ClockIn period		200	400	ns	1,3
TDCerror	ClockIn timing error			±0.5	ns	2
TDC1DC2	Difference in ClockIn for 2 linked devices			400	ppm	3
TDCr	ClockIn rise time			10	ns	4
TDCf	ClockIn fall time			8	ns	4

Notes

1 Measured between corresponding points on consecutive falling edges.

2 Variation of individual falling edges from their nominal times.

3 This value allows the use of 200ppm crystal oscillators for two devices connected together by a link.

4 Clock transitions must be monotonic within the range **VIH** to **VIL** (page 266).

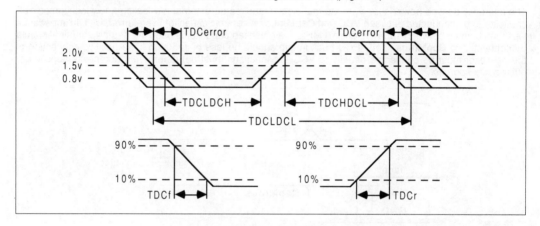

Figure 3.2: ClockIn timing

3.4 SeparateIQ

The IMS C011 link adaptor has two different modes of operation. Mode 1 is basically a link to peripheral adaptor, whilst Mode 2 interfaces between a link and a microprocessor bus system.

Mode 1 can be selected for one of two link speeds by connecting **SeparateIQ** to **VCC** (10 Mbits/sec) or to **ClockIn** (20 Mbits/sec).

Mode 2 is selected by connecting **SeparateIQ** to **GND**; in this mode 10 Mbits/sec or 20 Mbits/sec is selected by **LinkSpeed**. Link speeds are specified for a **ClockIn** frequency of 5 MHz.

In order to select the link speed, **SeparateIQ** may be changed dynamically providing the link is in a quiescent state and no input or output is required. **Reset** must be applied subsequent to the selection to initialise the device. If **ClockIn** is gated to achieve this, its skew must be limited to the value **TDCHSIQH** shown in table 3.2. The mode of operation (Mode 1, Mode 2) must not be changed dynamically.

3 System services

Table 3.2: SeparateIQ mode selection

SeparateIQ	Mode	Link Speed Mbits/sec
VCC	1	10
ClockIn	1	20
GND	2	10 or 20

Table 3.3: SeparateIQ

SYMBOL	PARAMETER	MIN	NOM	MAX	UNITS	NOTE
TDCHSIQH	Skew from ClockIn to ClockIn			20	ns	1

Notes

1 Skew between **ClockIn** arriving on the **ClockIn** pin and on the **SeparateIQ** pin.

3.5 Reset

The **Reset** pin can go high with **VCC**, but must at no time exceed the maximum specified voltage for **VIH**. After **VCC** is valid **ClockIn** should be running for a minimum period **TDCVRL** before the end of **Reset**. **LinkIn** must be held low during **Reset**.

Reset initialises the IMS C011 to the following state: **LinkOut** is held low; the control outputs (**IAck** and **QValid** in Mode 1, **InputInt** and **OutputInt** in Mode 2) are held low; interrupts (Mode 2) are disabled; the states of **Q0-7** in Mode 1 are unspecified; **D0-7** in Mode 2 are high impedance.

Table 3.4: Reset

SYMBOL	PARAMETER	MIN	NOM	MAX	UNITS	NOTE
TPVRH	Power valid before Reset	10			ms	
TRHRL	Reset pulse width high	8			ClockIn	1
TDCVRL	ClockIn running before Reset end	10			ms	2

Notes

1 Full periods of **ClockIn** TDCLDCL required.

2 At power-on reset.

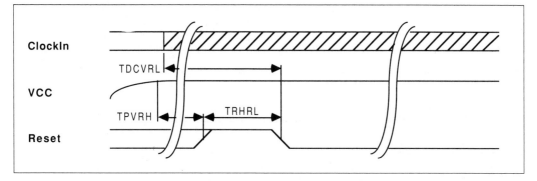

Figure 3.3: Reset timing

4 Links

INMOS bi-directional serial links provide synchronized communication between INMOS products and with the outside world. Each link comprises an input channel and output channel. A link between two devices is implemented by connecting a link interface on one device to a link interface on the other device. Every byte of data sent on a link is acknowledged on the input of the same link, thus each signal line carries both data and control information.

The quiescent state of a link output is low. Each data byte is transmitted as a high start bit followed by a one bit followed by eight data bits followed by a low stop bit. The least significant bit of data is transmitted first. After transmitting a data byte the sender waits for the acknowledge, which consists of a high start bit followed by a zero bit. The acknowledge signifies both that a process was able to receive the acknowledged data byte and that the receiving link is able to receive another byte.

Links are not synchronised with **ClockIn** and are insensitive to its phase. Thus links from independently clocked systems may communicate, providing only that the clocks are nominally identical and within specification.

Links are TTL compatible and intended to be used in electrically quiet environments, between devices on a single printed circuit board or between two boards via a backplane. Direct connection may be made between devices separated by a distance of less than 300 millimetres. For longer distances a matched 100 Ohm transmission line should be used with series matching resistors **RM**. When this is done the line delay should be less than 0.4 bit time to ensure that the reflection returns before the next data bit is sent.

Buffers may be used for very long transmissions. If so, their overall propagation delay should be stable within the skew tolerance of the link, although the absolute value of the delay is immaterial.

The IMS C011 link supports the standard INMOS communication speed of 10 Mbits per second. In addition it can be used at 20 Mbits per second. Link speed can be selected in one of two ways. In Mode 1 it is altered by **SeparateIQ** (page 254). In Mode 2 it is selected by **LinkSpeed**; when the **LinkSpeed** pin is low, the link operates at the standard 10 Mbits/sec; when high it operates at 20 Mbits/sec.

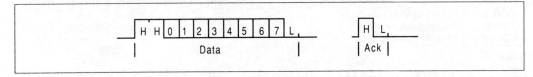

Figure 4.1: IMS C011 link data and acknowledge packets

4 Links

Table 4.1: Link

SYMBOL	PARAMETER		MIN	NOM	MAX	UNITS	NOTE
TJQr	LinkOut rise time				20	ns	
TJQf	LinkOut fall time				10	ns	
TJDr	LinkIn rise time				20	ns	
TJDf	LinkIn fall time				20	ns	
TJQJD	Buffered edge delay		0			ns	
TJBskew	Variation in TJQJD	20 Mbits/s			3	ns	1
		10 Mbits/s			10	ns	1
CLIZ	LinkIn capacitance	@ f=1MHz			7	pF	
CLL	LinkOut load capacitance				50	pF	
RM	Series resistor for 100Ω transmission line			56		ohms	

Notes

1 This is the variation in the total delay through buffers, transmission lines, differential receivers etc., caused by such things as short term variation in supply voltages and differences in delays for rising and falling edges.

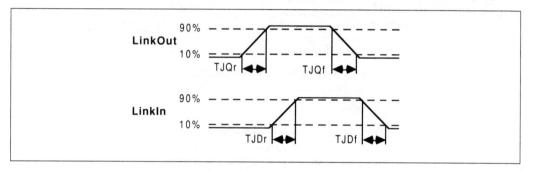

Figure 4.2: IMS C011 link timing

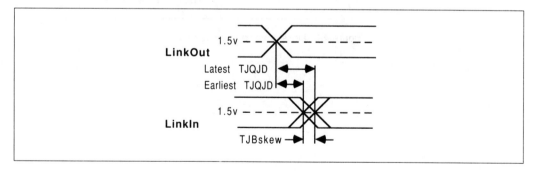

Figure 4.3: IMS C011 buffered link timing

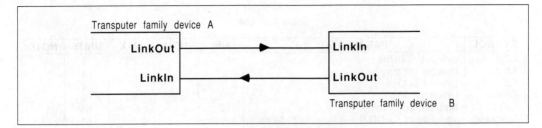

Figure 4.4: Links directly connected

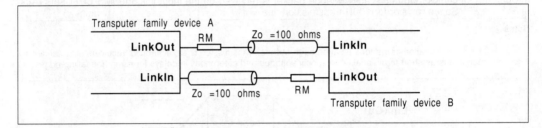

Figure 4.5: Links connected by transmission line

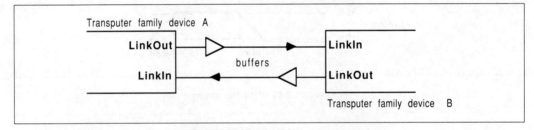

Figure 4.6: Links connected by buffers

5 Mode 1 parallel interface

In Mode 1 the IMS C011 link adaptor is configured as a parallel peripheral interface with handshake lines. Communication with a transputer family device is via the serial link. The parallel interface comprises an input port and an output port, both with handshake.

5.1 Input port

The eight bit parallel input port **I0-7** can be read by a transputer family device via the serial link. **IValid** and **IAck** provide a simple two-wire handshake for this port. When data is valid on **I0-7**, **IValid** is taken high by the peripheral device to commence the handshake. The link adaptor transmits data presented on **I0-7** out through the serial link. When the acknowledge packet is received on the input link, the IMS C011 sets **IAck** high. To complete the handshake, the peripheral device must return **IValid** low. The link adaptor will then set **IAck** low. New data should not be put onto **I0-7** until **IAck** is returned low.

Table 5.1: Mode 1 parallel data output

SYMBOL	PARAMETER	MIN	NOM	MAX	UNITS	NOTE
TLdVQvH	Start of link data to QValid	11.5			bits	1
TQdVQvH	Data setup	15			ns	2
TQvHQaH	QAck setup time from QValid high	0			ns	
TQaHQvL	QAck high to QValid low	1.8			bits	1
TQaHLaV	QAck high to Ack on link	0.8		2	bits	1,3
TQvLQaL	QAck hold after QValid low	0			ns	
TQvLQdX	Data hold	11			bits	1,4

Notes

1. Unit of measurement is one link data bit time; at 10 Mbits/s data link speed, one bit time is nominally 100 nS.

2. Where an existing data output bit is re-written with the same level there will be no glitch in the output level.

3. Maximum time assumes there is no data packet already on the link. Maximum time with data on the link is extended by 11 bits.

4. Data output remains valid until overwritten by new data.

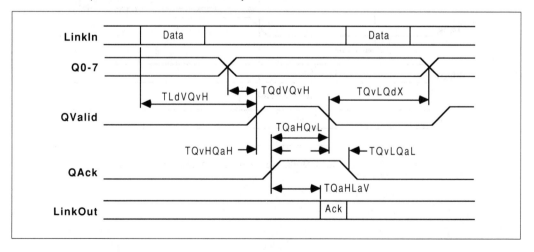

Figure 5.1: IMS C011 Mode 1 parallel data output from link adaptor

5.2 Output port

The eight bit parallel output port **Q0-7** can be controlled by a transputer family device via the serial link. **QValid** and **QAck** provide a simple two-wire handshake for this port.

A data packet received on the input link is presented on **Q0-7**; the link adaptor then takes **QValid** high to initiate the handshake. After reading data from **Q0-7**, the peripheral device sets **QAck** high. The IMS C011 will then send an acknowledge packet out of the serial link to indicate a completed transaction and set **QValid** low to complete the handshake.

Table 5.2: Mode 1 parallel data input

SYMBOL	PARAMETER	MIN	NOM	MAX	UNITS	NOTE
TIdVIvH	Data setup	5			ns	
TIvHLdV	IValid high to link data output	0.8		2	bits	1,2
TLaVIaH	Link acknowledge start to IAck high			3	bits	1
TIaHIdX	Data hold after IAck high	0			ns	
TIaHIvL	IValid hold after IAck high	0			ns	
TIvLIaL	IAck hold after IValid low	1		4	bits	1
TIaLIvH	Delay before next IValid high	0			ns	

Notes

1. Unit of measurement is one link data bit time; at 10 Mbits/s data link speed, one bit time is nominally 100 nS.

2. Maximum time assumes there is no acknowledge packet already on the link. Maximum time with acknowledge on the link is extended by 2 bits.

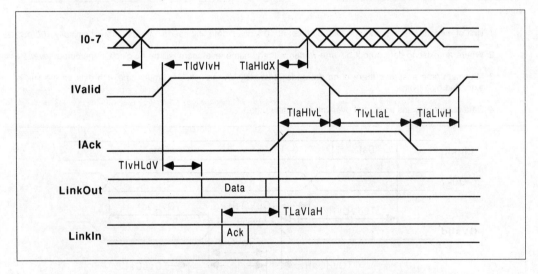

Figure 5.2: IMS C011 Mode 1 parallel data input to link adaptor

6 Mode 2 parallel interface

The IMS C011 provides an interface between a link and a microprocessor style bus. Operation of the link adaptor is controlled through the parallel interface bus lines **D0-7** by reading and writing various registers in the link adaptor. Registers are selected by **RS0-1** and **RnotW**, and the chip enabled with **notCS**.

For convenience of description, the device connected to the parallel side of the link adaptor is presumed to be a microprocessor, although this will not always be the case.

6.1 D0-7

Data is communicated between a microprocessor bus and the link adaptor via the bidirectional bus lines **D0-7**. The bus is high impedance unless the link adaptor chip is selected and the **RnotW** line is high. The bus is used by the microprocessor to access status and data registers.

6.2 notCS

The link adaptor chip is selected when **notCS** is low. Register selectors **RS0-1** and **RnotW** must be valid before **notCS** goes low; **D0-7** must also be valid if writing to the chip (**RnotW** low). Data is read by the link adaptor on the rising edge of **notCS**.

6.3 RnotW

RnotW, in conjunction with **notCS**, selects the link adaptor registers for read or write mode. When **RnotW** is high, the contents of an addressed register appear on the data bus **D0-7**; when **RnotW** is low the data on **D0-7** is written into the addressed register. The state of **RnotW** is latched into the link adaptor by **notCS** going low; it may be changed before **notCS** returns high, within the timing restrictions given.

6.4 RS0-1

One of four registers is selected by **RS0-1**. A register is addressed by setting up **RS0-1** and then taking **notCS** low; the state of **RnotW** when **notCS** goes low determines whether the register will be read or written. The state of **RS0-1** is latched into the link adaptor by **notCS** going low; it may be changed before **notCS** returns high, within the timing restrictions given. The register set comprises a read-only data input register, a write-only data output register and a read/write status register for each.

Table 6.1: IMS C011 Mode 2 register selection

RS1	RS0	RnotW	Register
0	0	1	Read data
0	0	0	Invalid
0	1	0	Invalid
0	1	0	Write data
1	0	1	Read input status
1	0	0	Write input status
1	1	1	Read output status
1	1	0	Write output status

6.4.1 Input Data Register

This register holds the last data packet received from the serial link. It never contains acknowledge packets. It contains valid data only whilst the *data present* flag is set in the input status register. It cannot be assumed to contain valid data after it has been read; a double read may or may not return valid data on the second read. If *data present* is valid on a subsequent read it indicates new data is in the buffer. Writing to this register will have no effect.

Table 6.2: IMS C011 Mode 2 parallel interface control

SYMBOL	PARAMETER	MIN	NOM	MAX	UNITS	NOTE
TRSVCSL	Register select setup	5			ns	
TCSLRSX	Register select hold	5			ns	
TRWVCSL	Read/write strobe setup	5			ns	
TCSLRWX	Read/write strobe hold	5			ns	
TCSLCSH	Chip select active	50			ns	
TCSHCSL	Delay before re-assertion of chip select	50			ns	

Table 6.3: IMS C011 Mode 2 parallel interface read

SYMBOL	PARAMETER	MIN	NOM	MAX	UNITS	NOTE
TLdVIIH	Start of link data to InputInt high			13	bits	1
TCSLIIL	Chip select to InputInt low			30	ns	
TCSLDrX	Chip select to bus active	5			ns	
TCSLDrV	Chip select to data valid			40	ns	
TCSHDrZ	Chip select high to bus tristate			25	ns	
TCSHDrX	Data hold after chip select high	5			ns	
TCSHLaV	Chip de-select to start of Ack	0.8		2	bits	1,2

Notes

1 Unit of measurement is one link data bit time; at 10 Mbits/s data link speed, one bit time is nominally 100 nS.

2 Maximum time assumes there is no data packet already on the link. Maximum time with data on the link is extended by 11 bits.

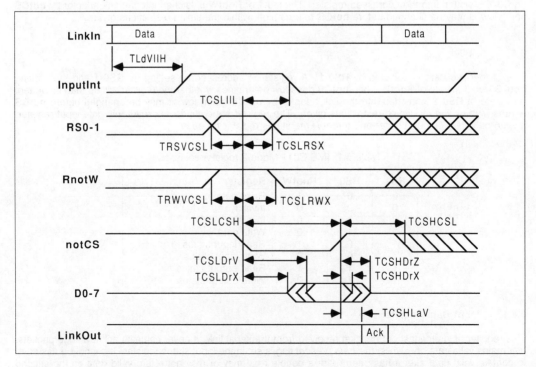

Figure 6.1: IMS C011 Mode 2 read parallel data from link adaptor

6 Mode 2 parallel interface

Table 6.4: IMS C011 Mode 2 parallel interface write

SYMBOL	PARAMETER	MIN	NOM	MAX	UNITS	NOTE
TCSHDwV	Data setup	15			ns	
TCSHDwX	Data hold	5			ns	
TCSLOIL	Chip select to OutputInt low			30	ns	
TCSHLdV	Chip select high to start of link data	0.8		2	bits	1,2
TLaVOIH	Start of link Ack to OutputInt high			3	bits	1,3
TLdVOIH	Start of link data to OutputInt high			13	bits	1,3

Notes

1 Unit of measurement is one link data bit time; at 10 Mbits/s data link speed, one bit time is nominally 100 nS.

2 Maximum time assumes there is no acknowledge packet already on the link. Maximum time with acknowledge on the link is extended by 2 bits.

3 Both data transmission and the returned acknowledge must be completed before **OutputInt** can go high.

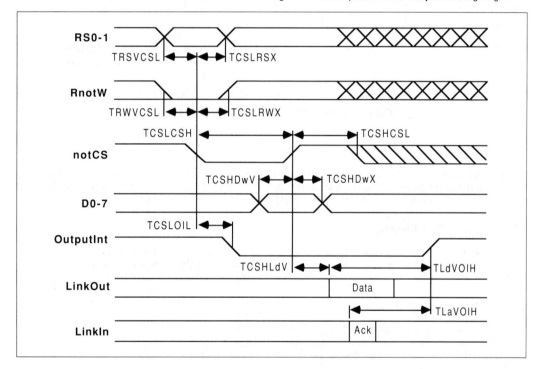

Figure 6.2: IMS C011 Mode 2 write parallel data to link adaptor

6.4.2 Input Status Register

This register contains the *data present* flag and the *interrupt enable* control bit for **InputInt**. The *data present* flag is set to indicate that data in the data input buffer is valid. It is reset low only when the data input buffer is read, or by **Reset**. When writing to this register, the *data present* bit must be written as zero.

The *interrupt enable* bit can be set and reset by writing to the status register with this bit high or low respectively. When the *interrupt enable* and *data present* flags are both high, the **InputInt** output will be high (page 264). Resetting *interrupt enable* will take **InputInt** low; setting it again before reading the data input register will set **InputInt** high again. The *interrupt enable* bit can be read to determine its status.

When writing to this register, bits 2-7 must be written as zero; this ensures that they will be zero when the register is read. Failure to write zeroes to these bits may result in undefined data being returned by these bits during a status register read.

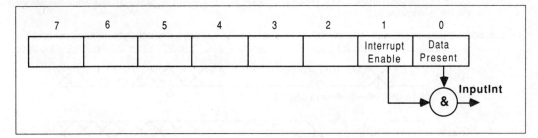

Figure 6.3: IMS C011 Mode 2 input status register

6.5 InputInt

The **InputInt** output is set high to indicate that a data packet has been received from the serial link. It is inhibited from going high when the *interrupt enable* bit in the input status register is low (page 264). **InputInt** is reset low when data is read from the input data register (page 261) and by **Reset** (page 255).

6.5.1 Output Data Register

Data written to this link adaptor register is transmitted out of the serial link as a data packet. Data should only be written to this register when the *output ready* bit in the output status register is high, otherwise data already being transmitted may be corrupted. Reading this register will result in undefined data being read.

6.5.2 Output Status Register

This register contains the *output ready* flag and the *interrupt enable* control bit for **OutputInt**. The *output ready* flag is set to indicate that the data output buffer is empty. It is reset low only when data is written to the data output buffer; it is set high by **Reset**. When writing to this register, the *output ready* bit must be written as zero.

The *interrupt enable* bit can be set and reset by writing to the status register with this bit high or low respectively. When the *interrupt enable* and *output ready* flags are both high, the **OutputInt** output will be high (page 265). Resetting *interrupt enable* will take **OutputInt** low; setting it again whilst the data output register is empty will set **OutputInt** high again. The *interrupt enable* bit can be read to determine it's status.

When writing to this register, bits 2-7 must be written as zero; this ensures that they will be zero when the register is read. Failure to write zeroes to these bits may result in undefined data being returned by these bits during a status register read.

6 Mode 2 parallel interface

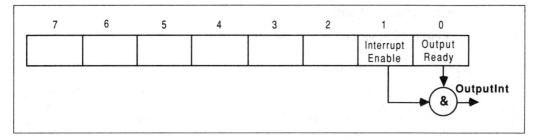

Figure 6.4: IMS C011 Mode 2 output status register

6.6 OutputInt

The **OutputInt** output is set high to indicate that the link is free to receive data from the microprocessor for transmission as a data packet out of the serial link. It is inhibited from going high when the *interrupt enable* bit in the output status register is low (page 264). **OutputInt** is reset low when data is written to the data output register (page 264); it is set high by **Reset** (page 255).

6.7 Data read

A data packet received on the input link sets the *data present* flag in the input status register. If the *interrupt enable* bit in the status register is set, the **InputInt** output pin will be set high. The microprocessor will either respond to the interrupt (if the *interrupt enable* bit is set) or will periodically read the input status register until the *data present* bit is high.

When data is available from the link, the microprocessor reads the data packet from the data input register. This will reset the *data present* flag and cause the link adaptor to transmit an acknowledge packet out of the serial link output. **InputInt** is automatically reset by reading the data input register; it is not necessary to read or write the input status register.

6.8 Data write

When the data output buffer is empty the *output ready* flag in the output status register is set high. If the *interrupt enable* bit in the status register is set, the **OutputInt** output pin will also be set high. The microprocessor will either respond to the interrupt (if the *interrupt enable* bit is set) or will periodically read the output status register until the *output ready* bit is high.

When the *output ready* flag is high, the microprocessor can write data to the data output buffer. This will result in the link adaptor resetting the *output ready* flag and commencing transmission of the data packet out of the serial link. The *output ready* status bit will remain low until an acknowledge packet is received by the input link. This will set the *output ready* flag high; if the *interrupt enable* bit is set, **OutputInt** will also be set high.

7 Electrical specifications

7.1 DC electrical characteristics

Table 7.1: Absolute maximum ratings

SYMBOL	PARAMETER	MIN	MAX	UNITS	NOTE
VCC	DC supply voltage	0	7.0	V	1,2,3
VI, VO	Voltage on input and output pins	-0.5	VCC+0.5	V	1,2,3
II	Input current		±25	mA	4
OSCT	Output short circuit time (one pin)		1	s	2
TS	Storage temperature	-65	150	°C	2
TA	Ambient temperature under bias	-55	125	°C	2
PDmax	Maximum allowable dissipation		600	mW	

Notes

1. All voltages are with respect to **GND**.

2. This is a stress rating only and functional operation of the device at these or any other conditions beyond those indicated in the operating sections of this specification is not implied. Stresses greater than those listed may cause permanent damage to the device. Exposure to absolute maximum rating conditions for extended periods may affect reliability.

3. This device contains circuitry to protect the inputs against damage caused by high static voltages or electrical fields. However, it is advised that normal precautions be taken to avoid application of any voltage higher than the absolute maximum rated voltages to this high impedance circuit. Unused inputs should be tied to an appropriate logic level such as **VCC** or **GND**.

4. The input current applies to any input or output pin and applies when the voltage on the pin is between **GND** and **VCC**.

Table 7.2: Operating conditions

SYMBOL	PARAMETER	MIN	MAX	UNITS	NOTE
VCC	DC supply voltage	4.75	5.25	V	1
VI, VO	Input or output voltage	0	VCC	V	1,2
CL	Load capacitance on any pin		50	pF	
TA	Operating temperature range	0	70	°C	3

Notes

1. All voltages are with respect to **GND**.

2. Excursions beyond the supplies are permitted but not recommended; see DC characteristics.

3. Air flow rate 400 linear ft/min transverse air flow.

7 Electrical specifications

Table 7.3: DC characteristics

SYMBOL	PARAMETER		MIN	MAX	UNITS	NOTE
VIH	High level input voltage		2.0	VCC+0.5	V	1,2
VIL	Low level input voltage		-0.5	0.8	V	1,2
II	Input current	@ GND<VI<VCC		±10	µA	1,2,6
				±200	µA	1,2,7
VOH	Output high voltage	@ IOH=2mA	VCC-1		V	1,2
VOL	Output low voltage	@ IOL=4mA		0.4	V	1,2
IOS	Output short circuit current	@ GND<VO<VCC		50	mA	1,2,4
				75	mA	1,2,5
IOZ	Tristate output current	@ GND<VO<VCC		±10	µA	1,2
PD	Power dissipation			120	mW	2,3
CIN	Input capacitance	@ f=1MHz		7	pF	
COZ	Output capacitance	@ f=1MHz		10	pF	

Notes

1 All voltages are with respect to **GND**.

2 Parameters measured at 4.75V<**VCC**<5.25V and 0°C<**TA**<70°C. Input clock frequency = 5MHz.

3 Power dissipation varies with output loading.

4 Current sourced from non-link outputs.

5 Current sourced from link outputs.

6 For inputs other than those in Note 7.

7 For pins 2, 3, 5, 6, 7, 9, 11, 13, 15, 16, 25.

7.2 Equivalent circuits

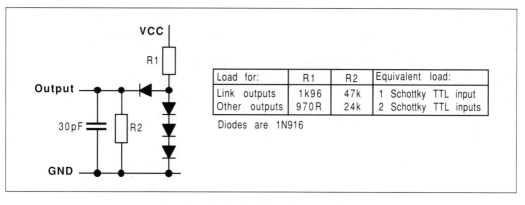

Figure 7.1: Load circuit for AC measurements

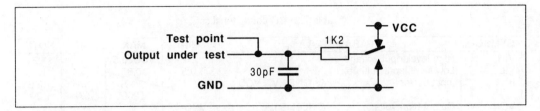

Figure 7.2: Tristate load circuit for AC measurements

7.3 AC timing characteristics

Table 7.4: Input, output edges

SYMBOL	PARAMETER	MIN	MAX	UNITS	NOTE
TDr	Input rising edges	2	20	ns	1,2
TDf	Input falling edges	2	20	ns	1,2
TQr	Output rising edges		25	ns	1
TQf	Output falling edges		15	ns	1
CSLaHZ	Chip select high to tristate		25	ns	
CSLaLZ	Chip select low to tristate		25	ns	

Notes

1 Non-link pins; see section on links.

2 All inputs except **ClockIn**; see section on **ClockIn**.

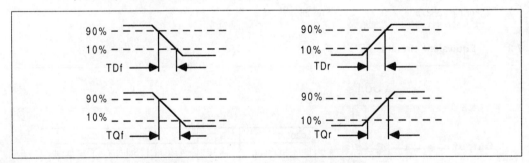

Figure 7.3: IMS C011 input and output edge timing

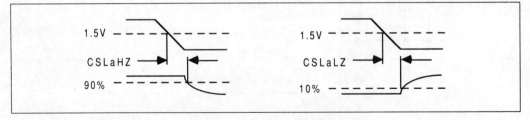

Figure 7.4: IMS C011 tristate timing relative to notCS

7 Electrical specifications

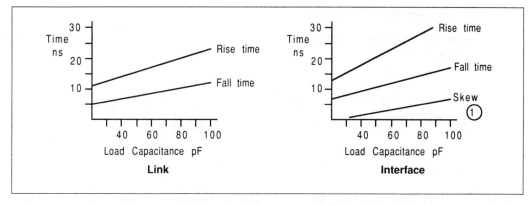

Notes

Figure 7.5: Typical rise/fall times

1. Skew is measured between **notCS** with a standard load (2 Schottky TTL inputs and 30pF) and **notCS** with a load of 2 Schottky TTL inputs and varying capacitance.

7.4 Power rating

Internal power dissipation P_{INT} of transputer and peripheral chips depends on **VCC**, as shown in figure 7.6. P_{INT} is substantially independent of temperature.

Total power dissipation P_D of the chip is

$$P_D = P_{INT} + P_{IO}$$

where P_{IO} is the power dissipation in the input and output pins; this is application dependent.

Internal working temperature T_J of the chip is

$$T_J = T_A + \theta J_A * P_D$$

where T_A is the external ambient temperature in °C and θJ_A is the junction-to-ambient thermal resistance in °C/W. θJ_A for each package is given in the Packaging Specifications section.

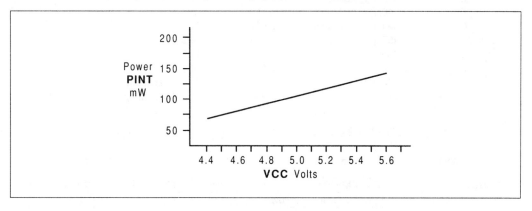

Figure 7.6: IMS C011 internal power dissipation vs VCC

8 Package specifications

8.1 28 pin plastic dual-in-line package

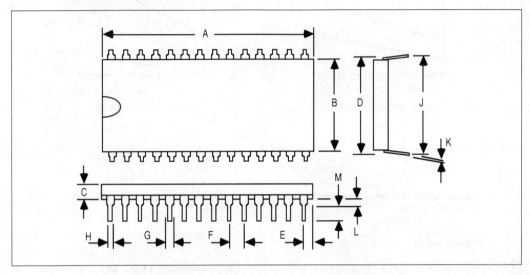

Figure 8.1: 28 pin plastic dual-in-line package dimensions

Table 8.1: 28 pin plastic dual-in-line package dimensions

DIM	Millimetres		Inches		Notes
	NOM	TOL	NOM	TOL	
A	36.830	+0.508	1.450	+0.020	
		−0.254		−0.010	
B	13.716	±0.051	0.540	±0.002	
C	3.810	±0.254	0.150	±0.010	
D	15.240	±0.076	0.600	±0.003	
E	1.905	±0.051	0.075	±0.002	
F	2.540	±0.051	0.100	±0.002	
G	1.524	±0.051	0.060	±0.002	
H	0.457	±0.051	0.018	±0.002	
J	16.256	0.508	0.640	0.020	
K	0.254	±0.025	0.010	±0.001	
L	0.58		0.020		Minimum
M	3.429		0.135		Maximum

Package weight is approximately 4 grams

Table 8.2: 28 pin plastic dual-in-line package junction to ambient thermal resistance

SYMBOL	PARAMETER	MIN	NOM	MAX	UNITS	NOTE
θ_{JA}	At 400 linear ft/min transverse air flow		110		°C/W	

8 Package specifications

8.2 28 pin ceramic dual-in-line package

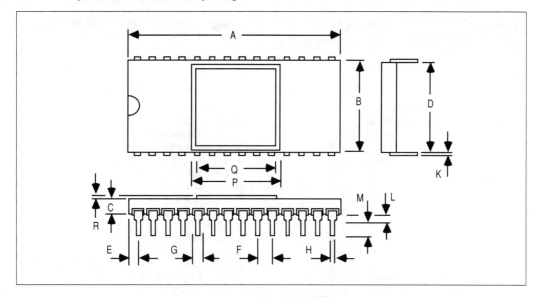

Figure 8.2: 28 pin ceramic dual-in-line package dimensions

Table 8.3: 28 pin ceramic dual-in-line package dimensions

DIM	Millimetres		Inches		Notes
	NOM	TOL	NOM	TOL	
A	35.560	±0.406	1.400	±0.016	
B	15.113	±0.254	0.595	±0.010	
C	2.159	±0.254	0.085	±0.010	
D	15.113	+0.762 / -0.000	0.595	+0.030 / -0.000	
E	2.032		0.080		Maximum
F	2.540	±0.127	0.100	±0.005	
G	1.270	±0.152	0.050	±0.006	
H	0.457	+0.102 / -0.051	0.018	+0.004 / -0.002	
K	0.254	+0.076 / -0.025	0.010	+0.003 / -0.001	
L	1.270	±0.381	0.050	±0.015	
M	3.683	±0.508	0.145	±0.020	
P	13.208	±0.229	0.520	±0.009	Square
Q	12.827	±0.127	0.505	±0.005	Square
R	0.3175	±0.0635	0.0125	±0.0025	

Package weight is approximately 5 grams

Table 8.4: 28 pin ceramic dual-in-line package junction to ambient thermal resistance

SYMBOL	PARAMETER	MIN	NOM	MAX	UNITS	NOTE
θJA	At 400 linear ft/min transverse air flow		60		°C/W	

8.3 Pinout

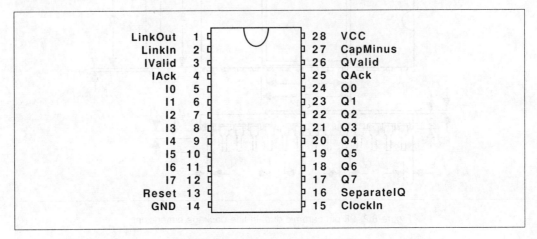

Figure 8.3: IMS C011 Mode 1 pinout

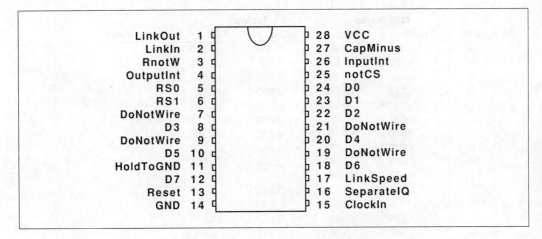

Figure 8.4: IMS C011 Mode 2 pinout

inmos

Chapter 9

IMS C012 engineering data

1 Introduction

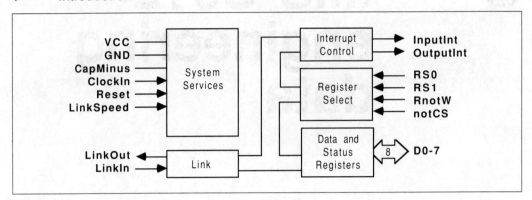

Figure 1.1: IMS C012 block diagram

1 Introduction

The INMOS communication link is a high speed system interconnect which provides full duplex communication between members of the INMOS transputer family, according to the INMOS serial link protocol. The IMS C012, a member of this family, provides for full duplex transputer link communication with standard microprocessor and sub-system architectures, by converting bi-directional serial link data into parallel data streams.

All INMOS products which use communication links, regardless of device type, support a standard communications frequency of 10 Mbits/sec; most products also support 20 Mbits/sec. Products of different type or performance can, therefore, be interconnected directly and future systems will be able to communicate directly with those of today. The IMS C012 link will run at either the standard speed of 10 Mbits/sec or at the higher speed of 20 Mbit/sec. Data reception is asynchronous, allowing communication to be independent of clock phase.

The IMS C012 provides an interface between an INMOS serial link and a microprocessor system bus. Status and data registers for both input and output ports can be accessed across the byte-wide bi-directional interface. Two interrupt outputs are provided, one to indicate input data available and one for output buffer empty.

2 Pin designations

Table 2.1: IMS C012 services and link

Pin	In/Out	Function
VCC, GND		Power supply and return
CapMinus		External capacitor for internal clock power supply
ClockIn	in	Input clock
Reset	in	System reset
LinkIn	in	Serial data input channel
LinkOut	out	Serial data output channel

Table 2.2: IMS C012 parallel interface

Pin	In/Out	Function
D0-7	in/out	Bi-directional data bus
notCS	in	Chip select
RS0-1	in	Register select
RnotW	in	Read/write control signal
InputInt	out	Interrupt on link receive buffer full
OutputInt	out	Interrupt on link transmit buffer empty
LinkSpeed	in	Select link speed as 10 or 20 Mbits/sec
HoldToGND		Must be connected to **GND**

Signal names are prefixed by **not** if they are active low, otherwise they are active high.
Pinout details for various packages are given on page 292.

3 System services

System services include all the necessary logic to start up and maintain the IMS C012.

3.1 Power

Power is supplied to the device via the **VCC** and **GND** pins. The supply must be decoupled close to the chip by at least one 100nF low inductance (e.g. ceramic) capacitor between **VCC** and **GND**. Four layer boards are recommended; if two layer boards are used, extra care should be taken in decoupling.

AC noise between **VCC** and **GND** must be kept below 200 mV peak to peak at all frequencies above 100 KHz. AC noise between **VCC** and the ground reference of load capacitances must be kept below 200 mV peak to peak at all frequencies above 30 MHz. Input voltages must not exceed specification with respect to **VCC** and **GND**, even during power-up and power-down ramping, otherwise *latchup* can occur. CMOS devices can be permanently damaged by excessive periods of latchup.

3.2 CapMinus

The internally derived power supply for internal clocks requires an external low leakage, low inductance 1µF capacitor to be connected between **VCC** and **CapMinus**. A ceramic capacitor is preferred, with an impedance less than 3 ohms between 100 KHz and 10 MHz. If a polarised capacitor is used the negative terminal should be connected to **CapMinus**. Total PCB track length should be less than 50mm. The positive connection of the capacitor must be connected directly to **VCC**. Connections must not otherwise touch power supplies or other noise sources.

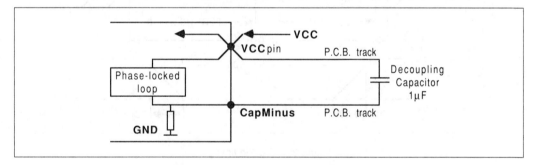

Figure 3.1: Recommended PLL decoupling

3.3 ClockIn

Transputer family components use a standard clock frequency, supplied by the user on the **ClockIn** input. The nominal frequency of this clock for all transputer family components is 5MHz, regardless of device type, transputer word length or processor cycle time. High frequency internal clocks are derived from **ClockIn**, simplifying system design and avoiding problems of distributing high speed clocks externally.

A number of transputer family devices may be connected to a common clock, or may have individual clocks providing each one meets the specified stability criteria. In a multi-clock system the relative phasing of **ClockIn** clocks is not important, due to the asynchronous nature of the links. Mark/space ratio is unimportant provided the specified limits of **ClockIn** pulse widths are met.

Oscillator stability is important. **ClockIn** must be derived from a crystal oscillator; RC oscillators are not sufficiently stable. **ClockIn** must not be distributed through a long chain of buffers. Clock edges must be monotonic and remain within the specified voltage and time limits.

Table 3.1: Input clock

SYMBOL	PARAMETER	MIN	NOM	MAX	UNITS	NOTE
TDCLDCH	ClockIn pulse width low	40			ns	
TDCHDCL	ClockIn pulse width high	40			ns	
TDCLDCL	ClockIn period		200	400	ns	1,3
TDCerror	ClockIn timing error			±0.5	ns	2
TDC1DC2	Difference in ClockIn for 2 linked devices			400	ppm	3
TDCr	ClockIn rise time			10	ns	4
TDCf	ClockIn fall time			8	ns	4

Notes

1 Measured between corresponding points on consecutive falling edges.

2 Variation of individual falling edges from their nominal times.

3 This value allows the use of 200ppm crystal oscillators for two devices connected together by a link.

4 Clock transitions must be monotonic within the range **VIH** to **VIL** (page 288).

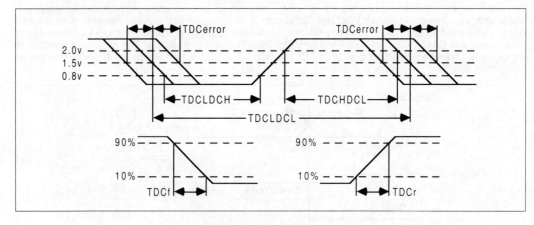

Figure 3.2: ClockIn timing

3 System services

3.4 Reset

The **Reset** pin can go high with **VCC**, but must at no time exceed the maximum specified voltage for **VIH**. After **VCC** is valid **ClockIn** should be running for a minimum period **TDCVRL** before the end of **Reset**. **LinkIn** must be held low during **Reset**.

Reset initialises the IMS C012 to the following state: **LinkOut** is held low; the interrupt outputs **InputInt** and **OutputInt** are held low; interrupts are disabled; **D0-7** are high impedance.

Table 3.2: Reset

SYMBOL	PARAMETER	MIN	NOM	MAX	UNITS	NOTE
TPVRH	Power valid before Reset	10			ms	
TRHRL	Reset pulse width high	8			ClockIn	1
TDCVRL	ClockIn running before Reset end	10			ms	2

Notes

1 Full periods of **ClockIn TDCLDCL** required.

2 At power-on reset.

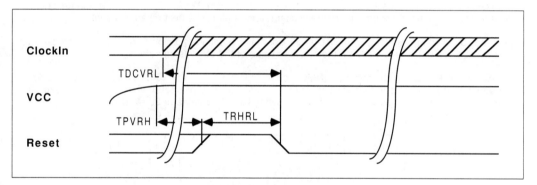

Figure 3.3: Reset timing

4   Links

INMOS bi-directional serial links provide synchronized communication between INMOS products and with the outside world. Each link comprises an input channel and output channel. A link between two devices is implemented by connecting a link interface on one device to a link interface on the other device. Every byte of data sent on a link is acknowledged on the input of the same link, thus each signal line carries both data and control information.

The quiescent state of a link output is low. Each data byte is transmitted as a high start bit followed by a one bit followed by eight data bits followed by a low stop bit. The least significant bit of data is transmitted first. After transmitting a data byte the sender waits for the acknowledge, which consists of a high start bit followed by a zero bit. The acknowledge signifies both that a process was able to receive the acknowledged data byte and that the receiving link is able to receive another byte.

Links are not synchronised with **ClockIn** and are insensitive to its phase. Thus links from independently clocked systems may communicate, providing only that the clocks are nominally identical and within specification.

Links are TTL compatible and intended to be used in electrically quiet environments, between devices on a single printed circuit board or between two boards via a backplane. Direct connection may be made between devices separated by a distance of less than 300 millimetres. For longer distances a matched 100 Ohm transmission line should be used with series matching resistors **RM**. When this is done the line delay should be less than 0.4 bit time to ensure that the reflection returns before the next data bit is sent.

Buffers may be used for very long transmissions. If so, their overall propagation delay should be stable within the skew tolerance of the link, although the absolute value of the delay is immaterial.

The IMS C012 link supports the standard INMOS communication speed of 10 Mbits per second. In addition it can be used at 20 Mbits per second. Link speed is selected by **LinkSpeed**; when the **LinkSpeed** pin is low, the link operates at the standard 10 Mbits/sec; when high it operates at 20 Mbits/sec.

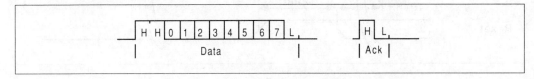

Figure 4.1: IMS C012 link data and acknowledge packets

4 Links

Table 4.1: Link

SYMBOL	PARAMETER		MIN	NOM	MAX	UNITS	NOTE
TJQr	LinkOut rise time				20	ns	
TJQf	LinkOut fall time				10	ns	
TJDr	LinkIn rise time				20	ns	
TJDf	LinkIn fall time				20	ns	
TJQJD	Buffered edge delay		0			ns	
TJBskew	Variation in TJQJD	20 Mbits/s			3	ns	1
		10 Mbits/s			10	ns	1
CLIZ	LinkIn capacitance	@ f=1MHz			7	pF	
CLL	LinkOut load capacitance				50	pF	
RM	Series resistor for 100Ω transmission line			56		ohms	

Notes

1 This is the variation in the total delay through buffers, transmission lines, differential receivers etc., caused by such things as short term variation in supply voltages and differences in delays for rising and falling edges.

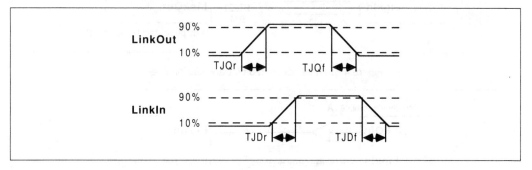

Figure 4.2: IMS C012 link timing

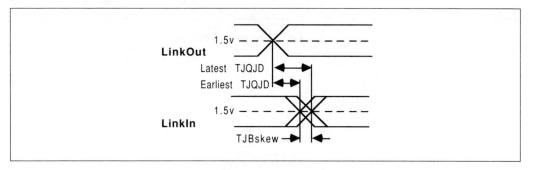

Figure 4.3: IMS C012 buffered link timing

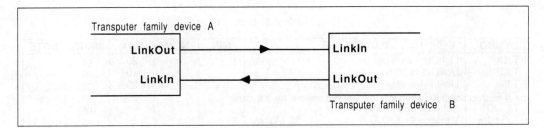

Figure 4.4: Links directly connected

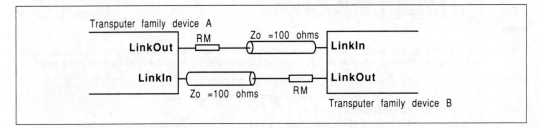

Figure 4.5: Links connected by transmission line

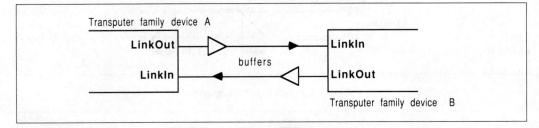

Figure 4.6: Links connected by buffers

5 Parallel interface

The IMS C012 provides an interface between a link and a microprocessor style bus. Operation of the link adaptor is controlled through the parallel interface bus lines **D0-7** by reading and writing various registers in the link adaptor. Registers are selected by **RS0-1** and **RnotW**, and the chip enabled with **notCS**.

For convenience of description, the device connected to the parallel side of the link adaptor is presumed to be a microprocessor, although this will not always be the case.

5.1 D0-7

Data is communicated between a microprocessor bus and the link adaptor via the bidirectional bus lines **D0-7**. The bus is high impedance unless the link adaptor chip is selected and the **RnotW** line is high. The bus is used by the microprocessor to access status and data registers.

5.2 notCS

The link adaptor chip is selected when **notCS** is low. Register selectors **RS0-1** and **RnotW** must be valid before **notCS** goes low; **D0-7** must also be valid if writing to the chip (**RnotW** low). Data is read by the link adaptor on the rising edge of **notCS**.

5.3 RnotW

RnotW, in conjunction with **notCS**, selects the link adaptor registers for read or write mode. When **RnotW** is high, the contents of an addressed register appear on the data bus **D0-7**; when **RnotW** is low the data on **D0-7** is written into the addressed register. The state of **RnotW** is latched into the link adaptor by **notCS** going low; it may be changed before **notCS** returns high, within the timing restrictions given.

5.4 RS0-1

One of four registers is selected by **RS0-1**. A register is addressed by setting up **RS0-1** and then taking **notCS** low; the state of **RnotW** when **notCS** goes low determines whether the register will be read or written. The state of **RS0-1** is latched into the link adaptor by **notCS** going low; it may be changed before **notCS** returns high, within the timing restrictions given. The register set comprises a read-only data input register, a write-only data output register and a read/write status register for each.

Table 5.1: IMS C012 register selection

RS1	RS0	RnotW	Register
0	0	1	Read data
0	0	0	Invalid
0	1	0	Invalid
0	1	0	Write data
1	0	1	Read input status
1	0	0	Write input status
1	1	1	Read output status
1	1	0	Write output status

5.4.1 Input Data Register

This register holds the last data packet received from the serial link. It never contains acknowledge packets. It contains valid data only whilst the *data present* flag is set in the input status register. It cannot be assumed to contain valid data after it has been read; a double read may or may not return valid data on the second read. If *data present* is valid on a subsequent read it indicates new data is in the buffer. Writing to this register will have no effect.

Table 5.2: IMS C012 parallel interface control

SYMBOL	PARAMETER	MIN	NOM	MAX	UNITS	NOTE
TRSVCSL	Register select setup	5			ns	
TCSLRSX	Register select hold	5			ns	
TRWVCSL	Read/write strobe setup	5			ns	
TCSLRWX	Read/write strobe hold	5			ns	
TCSLCSH	Chip select active	50			ns	
TCSHCSL	Delay before re-assertion of chip select	50			ns	

Table 5.3: IMS C012 parallel interface read

SYMBOL	PARAMETER	MIN	NOM	MAX	UNITS	NOTE
TLdVIIH	Start of link data to InputInt high			13	bits	1
TCSLIIL	Chip select to InputInt low			30	ns	
TCSLDrX	Chip select to bus active	5			ns	
TCSLDrV	Chip select to data valid			40	ns	
TCSHDrZ	Chip select high to bus tristate			25	ns	
TCSHDrX	Data hold after chip select high	5			ns	
TCSHLaV	Chip de-select to start of Ack	0.8		2	bits	1,2

Notes

1 Unit of measurement is one link data bit time; at 10 Mbits/s data link speed, one bit time is nominally 100 nS.

2 Maximum time assumes there is no data packet already on the link. Maximum time with data on the link is extended by 11 bits.

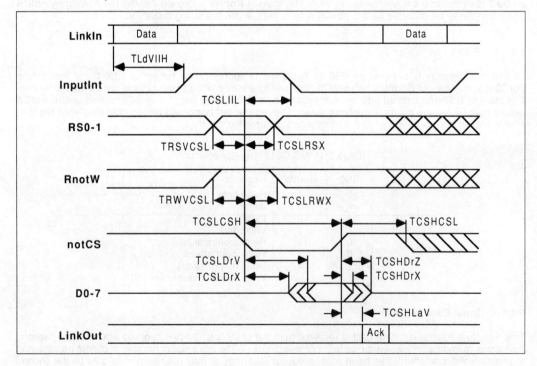

Figure 5.1: IMS C012 read parallel data from link adaptor

5 Parallel interface

Table 5.4: IMS C012 parallel interface write

SYMBOL	PARAMETER	MIN	NOM	MAX	UNITS	NOTE
TCSHDwV	Data setup	15			ns	
TCSHDwX	Data hold	5			ns	
TCSLOIL	Chip select to OutputInt low			30	ns	
TCSHLdV	Chip select high to start of link data	0.8		2	bits	1,2
TLaVOIH	Start of link Ack to OutputInt high			3	bits	1,3
TLdVOIH	Start of link data to OutputInt high			13	bits	1,3

Notes

1 Unit of measurement is one link data bit time; at 10 Mbits/s data link speed, one bit time is nominally 100 nS.

2 Maximum time assumes there is no acknowledge packet already on the link. Maximum time with acknowledge on the link is extended by 2 bits.

3 Both data transmission and the returned acknowledge must be completed before **OutputInt** can go high.

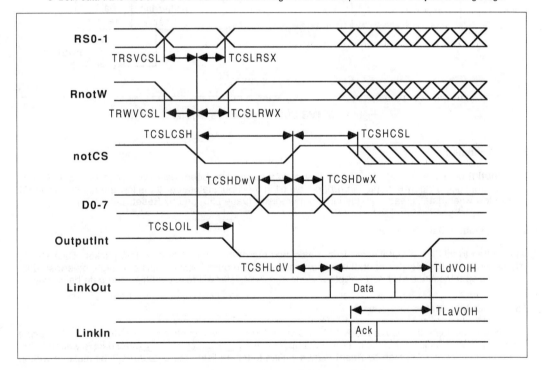

Figure 5.2: IMS C012 write parallel data to link adaptor

5.4.2 Input Status Register

This register contains the *data present* flag and the *interrupt enable* control bit for **InputInt**. The *data present* flag is set to indicate that data in the data input buffer is valid. It is reset low only when the data input buffer is read, or by **Reset**. When writing to this register, the *data present* bit must be written as zero.

The *interrupt enable* bit can be set and reset by writing to the status register with this bit high or low respectively. When the *interrupt enable* and *data present* flags are both high, the **InputInt** output will be high (page 286). Resetting *interrupt enable* will take **InputInt** low; setting it again before reading the data input register will set **InputInt** high again. The *interrupt enable* bit can be read to determine its status.

When writing to this register, bits 2-7 must be written as zero; this ensures that they will be zero when the register is read. Failure to write zeroes to these bits may result in undefined data being returned by these bits during a status register read.

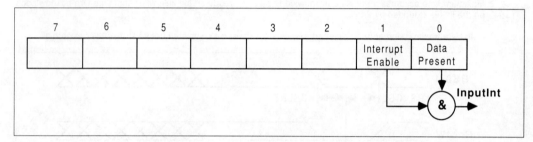

Figure 5.3: IMS C012 input status register

5.5 InputInt

The **InputInt** output is set high to indicate that a data packet has been received from the serial link. It is inhibited from going high when the *interrupt enable* bit in the input status register is low (page 286). **InputInt** is reset low when data is read from the input data register (page 283) and by **Reset** (page 279).

5.5.1 Output Data Register

Data written to this link adaptor register is transmitted out of the serial link as a data packet. Data should only be written to this register when the *output ready* bit in the output status register is high, otherwise data already being transmitted may be corrupted. Reading this register will result in undefined data being read.

5.5.2 Output Status Register

This register contains the *output ready* flag and the *interrupt enable* control bit for **OutputInt**. The *output ready* flag is set to indicate that the data output buffer is empty. It is reset low only when data is written to the data output buffer; it is set high by **Reset**. When writing to this register, the *output ready* bit must be written as zero.

The *interrupt enable* bit can be set and reset by writing to the status register with this bit high or low respectively. When the *interrupt enable* and *output ready* flags are both high, the **OutputInt** output will be high (page 287). Resetting *interrupt enable* will take **OutputInt** low; setting it again whilst the data output register is empty will set **OutputInt** high again. The *interrupt enable* bit can be read to determine it's status.

When writing to this register, bits 2-7 must be written as zero; this ensures that they will be zero when the register is read. Failure to write zeroes to these bits may result in undefined data being returned by these bits during a status register read.

5 Parallel interface

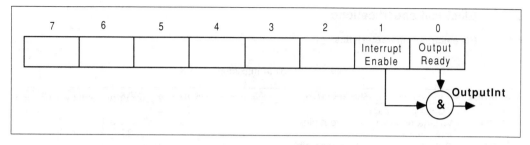

Figure 5.4: IMS C012 output status register

5.6 OutputInt

The **OutputInt** output is set high to indicate that the link is free to receive data from the microprocessor for transmission as a data packet out of the serial link. It is inhibited from going high when the *interrupt enable* bit in the output status register is low (page 286). **OutputInt** is reset low when data is written to the data output register (page 286); it is set high by **Reset** (page 279).

5.7 Data read

A data packet received on the input link sets the *data present* flag in the input status register. If the *interrupt enable* bit in the status register is set, the **InputInt** output pin will be set high. The microprocessor will either respond to the interrupt (if the *interrupt enable* bit is set) or will periodically read the input status register until the *data present* bit is high.

When data is available from the link, the microprocessor reads the data packet from the data input register. This will reset the *data present* flag and cause the link adaptor to transmit an acknowledge packet out of the serial link output. **InputInt** is automatically reset by reading the data input register; it is not necessary to read or write the input status register.

5.8 Data write

When the data output buffer is empty the *output ready* flag in the output status register is set high. If the *interrupt enable* bit in the status register is set, the **OutputInt** output pin will also be set high. The microprocessor will either respond to the interrupt (if the *interrupt enable* bit is set) or will periodically read the output status register until the *output ready* bit is high.

When the *output ready* flag is high, the microprocessor can write data to the data output buffer. This will result in the link adaptor resetting the *output ready* flag and commencing transmission of the data packet out of the serial link. The *output ready* status bit will remain low until an acknowledge packet is received by the input link. This will set the *output ready* flag high; if the *interrupt enable* bit is set, **OutputInt** will also be set high.

6 Electrical specifications

6.1 DC electrical characteristics

Table 6.1: Absolute maximum ratings

SYMBOL	PARAMETER	MIN	MAX	UNITS	NOTE
VCC	DC supply voltage	0	7.0	V	1,2,3
VI, VO	Voltage on input and output pins	-0.5	VCC+0.5	V	1,2,3
II	Input current		±25	mA	4
OSCT	Output short circuit time (one pin)		1	s	2
TS	Storage temperature	-65	150	°C	2
TA	Ambient temperature under bias	-55	125	°C	2
PDmax	Maximum allowable dissipation		600	mW	

Notes

1. All voltages are with respect to **GND**.

2. This is a stress rating only and functional operation of the device at these or any other conditions beyond those indicated in the operating sections of this specification is not implied. Stresses greater than those listed may cause permanent damage to the device. Exposure to absolute maximum rating conditions for extended periods may affect reliability.

3. This device contains circuitry to protect the inputs against damage caused by high static voltages or electrical fields. However, it is advised that normal precautions be taken to avoid application of any voltage higher than the absolute maximum rated voltages to this high impedance circuit. Unused inputs should be tied to an appropriate logic level such as **VCC** or **GND**.

4. The input current applies to any input or output pin and applies when the voltage on the pin is between **GND** and **VCC**.

Table 6.2: Operating conditions

SYMBOL	PARAMETER	MIN	MAX	UNITS	NOTE
VCC	DC supply voltage	4.75	5.25	V	1
VI, VO	Input or output voltage	0	VCC	V	1,2
CL	Load capacitance on any pin		50	pF	
TA	Operating temperature range	0	70	°C	3

Notes

1. All voltages are with respect to **GND**.

2. Excursions beyond the supplies are permitted but not recommended; see DC characteristics.

3. Air flow rate 400 linear ft/min transverse air flow.

6 Electrical specifications

Table 6.3: DC characteristics

SYMBOL	PARAMETER		MIN	MAX	UNITS	NOTE
VIH	High level input voltage		2.0	VCC+0.5	V	1,2
VIL	Low level input voltage		-0.5	0.8	V	1,2
II	Input current	@ GND<VI<VCC		±10	µA	1,2,6
				±200	µA	1,2,7
VOH	Output high voltage	@ IOH=2mA	VCC-1		V	1,2
VOL	Output low voltage	@ IOL=4mA		0.4	V	1,2
IOS	Output short circuit current	@ GND<VO<VCC		50	mA	1,2,4
				75	mA	1,2,5
IOZ	Tristate output current	@ GND<VO<VCC		±10	µA	1,2
PD	Power dissipation			120	mW	2,3
CIN	Input capacitance	@ f=1MHz		7	pF	
COZ	Output capacitance	@ f=1MHz		10	pF	

Notes

1. All voltages are with respect to **GND**.
2. Parameters measured at 4.75V<**VCC**<5.25V and 0°C<**TA**<70°C. Input clock frequency = 5MHz.
3. Power dissipation varies with output loading.
4. Current sourced from non-link outputs.
5. Current sourced from link outputs.
6. For inputs other than those in Note 7.
7. For pins 2, 3, 5, 6, 7, 9, 11, 13, 14, 21.

6.2 Equivalent circuits

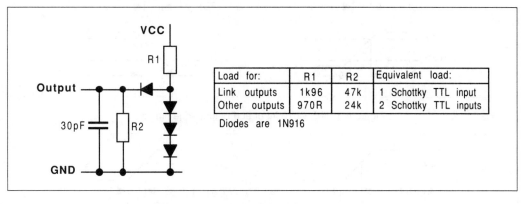

Figure 6.1: Load circuit for AC measurements

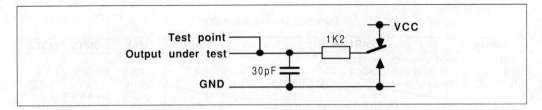

Figure 6.2: Tristate load circuit for AC measurements

6.3 AC timing characteristics

Table 6.4: Input, output edges

SYMBOL	PARAMETER	MIN	MAX	UNITS	NOTE
TDr	Input rising edges	2	20	ns	1,2
TDf	Input falling edges	2	20	ns	1,2
TQr	Output rising edges		25	ns	1
TQf	Output falling edges		15	ns	1
CSLaHZ	Chip select high to tristate		25	ns	
CSLaLZ	Chip select low to tristate		25	ns	

Notes

1 Non-link pins; see section on links.

2 All inputs except **ClockIn**; see section on **ClockIn**.

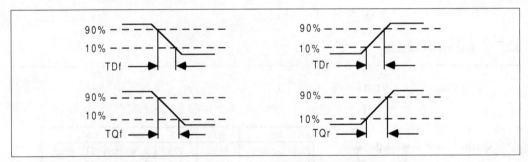

Figure 6.3: IMS C012 input and output edge timing

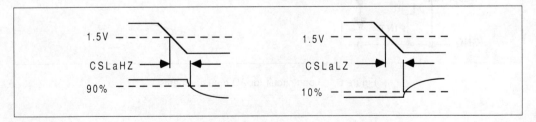

Figure 6.4: IMS C012 tristate timing relative to notCS

6 Electrical specifications

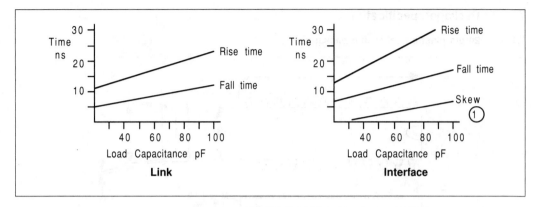

Figure 6.5: Typical rise/fall times

Notes

1 Skew is measured between **notCS** with a standard load (2 Schottky TTL inputs and 30pF) and **notCS** with a load of 2 Schottky TTL inputs and varying capacitance.

6.4 Power rating

Internal power dissipation P_{INT} of transputer and peripheral chips depends on **VCC**, as shown in figure 6.6. P_{INT} is substantially independent of temperature.

Total power dissipation P_D of the chip is

$$P_D = P_{INT} + P_{IO}$$

where P_{IO} is the power dissipation in the input and output pins; this is application dependent.

Internal working temperature T_J of the chip is

$$T_J = T_A + \theta J_A * P_D$$

where T_A is the external ambient temperature in °C and θJ_A is the junction-to-ambient thermal resistance in °C/W. θJ_A for each package is given in the Packaging Specifications section.

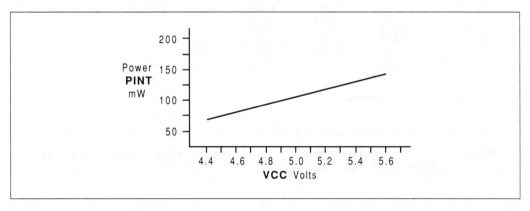

Figure 6.6: IMS C012 internal power dissipation vs VCC

7 Package specifications

7.1 24 pin plastic dual-in-line package

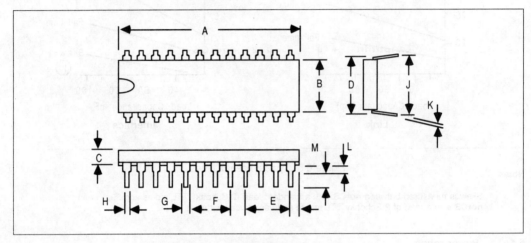

Figure 7.1: 24 pin plastic dual-in-line package dimensions

Table 7.1: 24 pin plastic dual-in-line package dimensions

DIM	Millimetres NOM	Millimetres TOL	Inches NOM	Inches TOL	Notes
A	31.242	+0.508 −0.254	1.230	+0.020 −0.010	
B	6.604	±0.127	0.260	±0.005	
C	3.302	±0.381	0.130	±0.015	
D	7.620	±0.127	0.300	±0.005	
E	1.651	±0.127	0.065	±0.005	
F	2.540	±0.127	0.100	±0.005	
G	1.524	±0.127	0.060	±0.005	
H	0.457	±0.127	0.018	±0.005	
J	8.382	±0.508	0.330	±0.020	
K	0.254	±0.025	0.010	±0.001	
L	0.508	±0.127	0.020	±0.005	
M	3.048		0.120		Minimum

Package weight is approximately 2 grams

Table 7.2: 24 pin plastic dual-in-line package junction to ambient thermal resistance

SYMBOL	PARAMETER	MIN	NOM	MAX	UNITS	NOTE
θJA	At 400 linear ft/min transverse air flow		115		°C/W	

7 Package specifications

7.2 Pinout

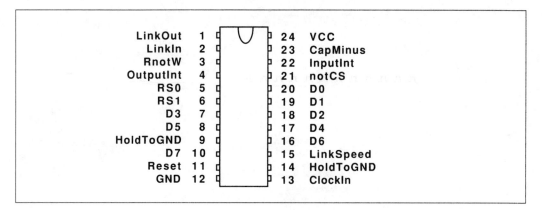

Figure 7.2: IMS C012 pinout

Appendix A

performance

A   Performance

The performance of the transputer is measured in terms of the number of bytes required for the program, and the number of (internal) processor cycles required to execute the program. The figures here relate to occam programs. For the same function, other languages should achieve approximately the same performance as occam. Unless otherwise stated, the IMS M212 characteristics are similar to those of the IMS T212.

With transputers incorporating an FPU, this type of performance calculation is straight forward when considering only integer data types. However, when floating point calculations using the **REAL32** and **REAL64** data types are present in the program, complications arise due to the concurrency inherent in the transputer's design whereby integer calculations can be overlapped with floating point calculations. A more comprehensive guide to the impact of this concurrency on transputer performance can be found in The Transputer Instruction Set - A Compiler Writers' Guide.

A.1   Performance overview

These figures are averages obtained from detailed simulation, and should be used only as an initial guide; they assume operands are of type **INT**. The abbreviations in table A.1 are used to represent the quantities indicated. In the replicator section of the table, figures in braces {} are not necessary if the number of replications is a compile time constant. To estimate performance, add together the time for the variable references and the time for the operation.

Table A.1: Key to performance table

np	number of component processes
ne	number of processes earlier in queue
r	1 if **INT** parameter or array parameter, 0 if not
ts	number of table entries (table size)
w	width of constant in nibbles
p	number of places to shift
Eg	expression used in a guard
Et	timer expression used in a guard
Tb	most significant bit set of multiplier ((-1) if the multiplier is 0)
Tbp	most significant bit set in a positive multiplier when counting from zero ((-1) if the multiplier is 0)
Tbc	most significant bit set in the two's complement of a negative multiplier
nsp	Number of scalar parameters in a procedure
nap	Number of array parameters in a procedure

Table A.2: Performance

		Size (bytes)	Time (cycles)
Names			
variables			
in expression		1.1+**r**	2.1+2(**r**)
assigned to or input to		1.1+**r**	1.1+(**r**)
in `PROC` or `FUNCTION` call,			
corresponding to an `INT` parameter		1.1+**r**	1.1+(**r**)
channels		1.1	2.1
Array Variables (for single dimension arrays)			
constant subscript		0	0
variable subscript		5.3	7.3
expression subscript		5.3	7.3
Declarations			
`CHAN OF` *protocol*		3.1	3.1
`[size]CHAN OF` *protocol*		9.4	2.2 + 20.2∗**size**
`PROC`		body+2	0
Primitives			
assignment		0	0
input		4	26.5
output		1	26
`STOP`		2	25
`SKIP`		0	0
Arithmetic operators			
+ −		1	1
∗	T2 only	2	23
∗	T4, T8	2	39
/	T2 only	2	24
/	T4, T8	2	40
`REM`	T2 only	2	22
`REM`	T4, T8	2	38
>> <<		2	3+**p**
Modulo Arithmetic operators			
`PLUS`		2	2
`MINUS`		1	1
`TIMES` (fast multiply)	T2, T4	1	4+**Tb**
`TIMES` (fast multiply, positive operand)	T8 only	1	4+**Tbp**
`TIMES` (fast multiply, negative operand)	T8 only	1	5+**Tbc**
Boolean operators			
`OR`		4	8
`AND` `NOT`		1	2
Comparison operators			
= constant		0	1
= variable		2	3
<> constant		1	3
<> variable		3	5
> <		1	2
>= <=		2	4
Bit operators			
/\ \/ >< ~		2	2

A Performance

Table A.3: Performance

	Size (bytes)	Time (cycles)
Expressions		
constant in expression	w	w
check if error	4	6
Timers		
timer input	2	3
timer **AFTER**		
if past time	2	4
with empty timer queue	2	31
non-empty timer queue	2	38+ne*9
ALT (timer)		
with empty timer queue	6	52
non-empty timer queue	6	59+ne*9
timer alt guard	8+2Eg+2Et	34+2Eg+2Et
Constructs		
SEQ	0	0
IF	1.3	1.4
if guard	3	4.3
ALT (non timer)	6	26
alt channel guard	10.2+2Eg	20+2Eg
skip alt guard	8+2Eg	10+2Eg
PAR	11.5+(np-1)*7.5	19.5+(np-1)*30.5
WHILE	4	12
Procedure or function call	3.5+(nsp-2)*1.1 +nap*2.3	16.5+(nsp-2)*1.1 +nap*2.3
Replicators		
replicated SEQ	7.3{+5.1}	(-3.8)+15.1*count{+7.1}
replicated IF	12.3{+5.1}	(-2.6)+19.4*count{+7.1}
replicated ALT	24.8{+10.2}	25.4+33.4*count{+14.2}
replicated timer ALT	24.8{+10.2}	62.4+33.4*count{+14.2}
replicated PAR	39.1{+5.1}	(-6.4)+70.9*count{+7.1}

A.2 Fast multiply, TIMES

Transputers have a fast integer multiplication instruction *product*. In the IMS T212 and IMS T414, if **Tb** is the position of the most significant bit set in the multiplier, then the time taken for a fast multiply is 4+**Tb**. The time taken for a multiplication by zero is 3 cycles. For example, if the multiplier is 1 the time taken is 4 cycles, if the multiplier is -1 (all bits set) the time taken is 19 cycles for the IMS T212 and 35 cycles for the IMS T414.

The IMS T800 has an improved performance for *product* over the IMS T212 and IMS T414. For a positive multiplier its execution time is 4+**Tbp** cycles, and for a negative multiplier 5+**Tbc** cycles (table A.1). The time taken for a multiplication by zero is 3 cycles.

Implementations of high level languages on the transputer may take advantage of this instruction. For example, the occam modulo arithmetic operator **TIMES** is implemented by the instruction and the right-hand operand is treated as the multiplier. The fast multiplication instruction is also used in high level language implementations for the multiplication implicit in multi-dimensional array access.

A.3 Arithmetic

A set of functions are provided within the development system to support the efficient implementation of multiple length and floating point arithmetic where relevant (in the IMS T800, floating point arithmetic is taken care of by the FPU). In table A.4 **n** gives the number of places shifted and all arguments and results are assumed to be local. Full details of these functions are provided in the occam reference manual, supplied as part of the development system and available as a separate publication.

When calculating the execution time of the predefined maths functions, no time needs to be added for calling overhead. These functions are compiled directly into special purpose instructions which are designed to support the efficient implementation of multiple length and floating point arithmetic.

Table A.4: Arithmetic performance

Function			Cycles	+ cycles for parameter access †
LONGADD			2	7
LONGSUM			3	8
LONGSUB			2	7
LONGDIFF			3	8
LONGPROD		T212 only	18	8
		T414, T800	34	8
LONGDIV		T212 only	20	8
		T414, T800	36	8
SHIFTRIGHT	(n<16)	T212 only	4+n	8
	(n>=16)	T212 only	n-11	8
	(n<32)	T414, T800	4+n	8
	(n>=32)	T414, T800	n-27	
SHIFTLEFT	(n<16)	T212 only	4+n	8
	(n>=16)	T212 only	n-11	8
	(n<32)	T414, T800	4+n	8
	(n>=32)	T414, T800	n-27	
NORMALISE	(n<16)	T212 only	n+6	7
	(n>=16)	T212 only	n-9	7
	(n=32)	T212 only	4	7
	(n<32)	T414, T800	n+6	7
	(n>=32)	T414, T800	n-25	
	(n=64)	T414, T800	4	
ASHIFTRIGHT			SHIFTRIGHT+2	5
ASHIFTLEFT			SHIFTLEFT+4	5
ROTATERIGHT			SHIFTRIGHT	7
ROTATELEFT			SHIFTLEFT	7
FRACMUL		T414, T800	LONGPROD+4	5

† Assuming local variables.

A.4 IMS T212, IMS T414 floating point operations

Floating point operations for IMS T212 and IMS T414 are provided by a run-time package. For the IMS T212 this requires approximately 2000 bytes of memory for the double length arithmetic operations, and 2500 bytes for the quadruple length arithmetic operations. For the IMS T212 it requires approximately 400 bytes of memory for the single length arithmetic operations, and 2500 bytes for the double length arithmetic operations. Table A.5 summarizes the estimated performance of the package.

A Performance

Table A.5: IMS T212, IMS T414 floating point operations performance

							Processor cycles			
							IMS T212		IMS T414	
							Typical	Worst	Typical	Worst
REAL32	+	−					530	705	230	300
	*						650	705	200	240
	/						1000	1410	245	280
	<	>	=	>=	<=	<>	60	60	60	60
REAL64	+	−					875	1190	565	700
	*						1490	1950	760	940
	/						2355	3255	1115	1420
	<	>	=	>=	<=	<>	60	60	60	60

A.5 IMS T800 floating point operations

All references to **REAL32** or **REAL64** operands within programs compiled for the IMS T800 should produce the following performance figures.

Table A.6: Floating point performance

	Size (bytes)	REAL32 Time (cycles)	REAL64 Time (cycles)
Names			
variables			
in expression	3.1	3	5
assigned to or input to	3.1	3	5
in **PROC** or **FUNCTION** call,			
corresponding to a **REAL**			
parameter	1.1+r	1.1+r	1.1+r
Arithmetic operators			
+ −	2	7	7
*	2	11	20
/	2	17	32
REM	11	19	34
Comparison operators			
=	2	4	4
<>	3	6	6
> <	2	5	5
>= <=	3	7	7
Conversions			
REAL32 to -	2		3
REAL64 to -	2	6	
To INT32 from -	5	9	9
To INT64 from -	18	32	32
INT32 to -	3	7	7
INT64 to -	14	24	22

A.5.1 IMS T800 floating point functions

These functions are provided by the development system. They are compiled directly into special purpose instructions designed to support the efficient implementation of some of the common mathematical functions of other languages. The functions provide **ABS** and **SQRT** for both **REAL32** and **REAL64** operand types.

Table A.7: IMS T800 floating point arithmetic performance

Function	Cycles	+ cycles for parameter access †	
		REAL32	REAL64
ABS	2	8	
SQRT	118	8	
DABS	2		12
DSQRT	244		12

† Assuming local variables.

A.5.2 IMS T800 special purpose functions and procedures

The functions and procedures given in tables A.9 and A.10 are provided by the development system to give access to the special instructions available on the IMS T800. Table A.8 shows the key to the table.

Table A.8: Key to special performance table

Tb	most significant bit set in the word counting from zero
n	number of words per row (consecutive memory locations)
r	number of rows in the two dimensional move
nr	number of bits to reverse

Table A.9: Special purpose functions performance

Function	Cycles	+ cycles for parameter access †
BITCOUNT	2+**Tb**	2
CRCBYTE	11	8
CRCWORD	35	8
BITREVNBIT	5+**nr**	4
BITREVWORD	36	2

† Assuming local variables.

Table A.10: Special purpose procedures performance

Procedure	Cycles	+ cycles for parameter access †
MOVE2D	8+(2n+23)∗**r**	8
DRAW2D	8+(2n+23)∗**r**	8
CLIP2D	8+(2n+23)∗**r**	8

† Assuming local variables.

A   Performance

A.6   Effect of external memory

Extra processor cycles may be needed when program and/or data are held in external memory, depending both on the operation being performed, and on the speed of the external memory. After a processor cycle which initiates a write to memory, the processor continues execution at full speed until at least the next memory access.

In the IMS M212 the on-chip ROM also requires extra processor cycles, as the access time of this memory is equivalent to three processor cycles for each word accessed.

Whilst a reasonable estimate may be made of the effect of external memory, the actual performance will depend upon the exact nature of the given sequence of operations.

External memory is characterized by the number of extra processor cycles per external memory cycle, denoted as e. The value of e for the IMS T212 and IMS M212 with no wait states is 1 and 4 respectively. For the IMS T414 and IMS T800, with the fastest external memory the value of e is 2; a typical value for a large external memory is 5. For the IMS M212, the value of e for the on chip ROM is 2.

If program is stored in external memory, and e has the value 2 or 3, then no extra cycles need be estimated for linear code sequences. For larger values of e, the number of extra cycles required for linear code sequences may be estimated at $(2e-1)/4$ per byte of program for IMS T212 and $(e-3)/4$ for the IMS T414 and IMS T800. A transfer of control may be estimated as requiring $e+3$ cycles.

These estimates may be refined for various constructs. In table A.11 n denotes the number of components in a construct. In the case of **IF**, the n'th conditional is the first to evaluate to **TRUE**, and the costs include the costs of the conditionals tested. The number of bytes in an array assignment or communication is denoted by b.

Table A.11: External memory performance

	IMS T212		IMS T414, IMS T800	
	Program off chip	Data off chip	Program off chip	Data off chip
Boolean expressions	e-1	0	e-2	0
IF	3en-1	en	3en-8	en
Replicated **IF**	6en+9e-12	(5e-2)n+6	(6e-4)n+7	(5e-2)n+8
Replicated **SEQ**	(4e-3)n+3e	(4e-2)n+3-e	(3e-3)n+2	(4e-2)n
PAR	4en	3en	(3e-1)n+8	3en+4
Replicated **PAR**	(17e-12)n+9	16en	(10e-8)n+8	16en-12
ALT	(4e-1)n+9e-4	(4e-1)n+9e-3	(2e-4)n+6e	(2e-2)n+10e-8
Array assignment and communication in one transputer	0	max (2e, eb)	0	max (2e, e(b/2))

For the IMS T212 and IMS T414, the effective rate of INMOS links is slowed down on output from external memory by e cycles per word output, and on input to external memory at 10 Mbits/sec by $e-6$ cycles per word if $e \geq 6$.

The following simulation results illustrate the effect of storing program and/or data in external memory. The results are normalized to 1 for both program and data on chip. The first program (Sieve of Erastosthenes) is an extreme case as it is dominated by small, data access intensive loops; it contains no concurrency, communication, or even multiplication or division. The second program is the pipeline algorithm for Newton Raphson square root computation.

Table A.12: IMS T212 external memory performance

	Program	e=1	e=2	e=3	e=4	On chip
Program off chip	1	1.2	1.4	1.8	2.1	1
	2	1.1	1.2	1.4	1.6	1
Data off chip	1	1.2	1.5	1.8	2.1	1
	2	1.1	1.3	1.4	1.6	1
Program and data off chip	1	1.4	1.9	2.5	3.0	1
	2	1.2	1.5	1.8	2.1	1

Table A.13: IMS T414, IMS T800 external memory performance

	Program	e=2	e=3	e=4	e=5	On chip
Program off chip	1	1.3	1.5	1.7	1.9	1
	2	1.1	1.2	1.2	1.3	1
Data off chip	1	1.5	1.8	2.1	2.3	1
	2	1.2	1.4	1.6	1.7	1
Program and data off chip	1	1.8	2.2	2.7	3.2	1
	2	1.3	1.6	1.8	2.0	1

Table A.14: IMS M212 external memory performance

	Program	e=4	On chip
Program off chip	1	2.1	1
	2	1.6	1
Data off chip	1	2.1	1
	2	1.6	1
Program and data off chip	1	3.0	1
	2	2.1	1

A.7 Interrupt latency

If the process is a high priority one and no other high priority process is running, the latency is as described in table A.15. The timings given are in full processor cycles **TPCLPCL**; the number of **Tm** states is also given where relevant. Maximum latency assumes all memory accesses are internal ones.

Table A.15: Interrupt latency

	Typical		Maximum	
	TPCLPCL	Tm	TPCLPCL	Tm
IMS T212, IMS M212	19		53	
IMS T414	19	38	58	116
IMS T800 with FPU in use	19	38	78	156
IMS T800 with FPU not in use	19	38	58	116

Appendix B

instruction set summary

B instruction set summary

B   Instruction set summary

The following tables give a comparison of the execution times of functions and operations for each transputer.

Where applicable, the FPU and processor operate concurrently, so the actual throughput of floating point instructions is better than that implied by simply adding up instruction times. For full details see The Transputer Instruction Set - A Compiler Writers' Guide.

The Processor Cycles column refers to the number of periods **TPCLPCL** taken by an instruction executing in internal memory. The number of cycles is given for the basic operation only; where relevant the time for the *prefix* function (one cycle) should be added. For a 20 MHz transputer one cycle is 50ns.

Some instruction times vary. Where a letter is included in the cycles or notes column it is interpreted from table B.1.

Table B.1: Instruction set interpretation

Ident	Interpretation
b	Bit number of the highest bit set in register *A*. Bit 0 is the least significant bit.
m	Bit number of the highest bit set in the absolute value of register *A*. Bit 0 is the least significant bit.
n	Number of places shifted.
n/a	Not applicable to this transputer.
p	Number of words per row.
r	Number of rows.
s	For 16 bit transputers this value is 16, for 32 bit transputers it is 32.
w	Number of words in the message. Part words are counted as full words. If the message is not word aligned the number of words is increased to include the part words at either end of the message.

Table B.2: Function codes

Mnemonic	Processor Cycles			Notes
	T212	T414	T800	
j	3	3	3	
ldlp	1	1	1	
pfix	1	1	1	
ldnl	2	2	2	
ldc	1	1	1	
ldnlp	1	1	1	
nfix	1	1	1	
ldl	2	2	2	
adc	1	1	1	
call	7	7	7	
cj	2	2	2	jump not taken
	4	4	4	jump taken
ajw	1	1	1	
eqc	2	2	2	
stl	1	1	1	
stnl	2	2	2	
opr	-	-	-	

Table B.3: Arithmetic/logical operation codes

Mnemonic	Processor Cycles			Notes
	T212	T414	T800	
and	1	1	1	
or	1	1	1	
xor	1	1	1	
not	1	1	1	
shl	n+2	n+2	n+2	
shr	n+2	n+2	n+2	
add	1	1	1	
sub	1	1	1	
mul	26	38	38	
fmul	n/a	35	35	no rounding
	n/a	40	40	rounding
div	23	39	39	
rem	21	37	37	
gt	2	2	2	
diff	1	1	1	
sum	1	1	1	
prod	b+4	b+4	b+4	positive register A
	b+4	b+4	m+5	negative register A

Table B.4: Long arithmetic operation codes

Mnemonic	Processor Cycles			Notes
	T212	T414	T800	
ladd	2	2	2	
lsub	2	2	2	
lsum	2	2	2	
ldiff	2	2	2	
lmul	17	33	33	
ldiv	19	35	35	
lshl	n+3	n+3	n+3	$n<s$
	n-12	n-28	n-28	$n \geq s$
lshr	n+3	n+3	n+3	$n<s$
	n-12	n-28	n-28	$n \geq s$
norm	n+5	n+5	n+5	$n<s$
	n-10	n-26	n-26	$n \geq s$
	3	3	3	$n=2*s$

Table B.5: General operation codes

Mnemonic	Processor Cycles			Notes
	T212	T414	T800	
rev	1	1	1	
xword	4	4	4	
cword	5	5	5	
xdble	2	2	2	
csngl	3	3	3	
mint	1	1	1	

B Instruction set summary

Table B.6: Indexing/array operation codes

Mnemonic	Processor Cycles			Notes
	T212	T414	T800	
bsub	1	1	1	
wsub	2	2	2	
wsubdb	n/a	n/a	3	
bcnt	2	2	2	
wcnt	4	5	5	
lb	5	5	5	
sb	4	4	4	
move	2w+8	2w+8	2w+8	

Table B.7: Timer handling operation codes

Mnemonic	Processor Cycles			Notes
	T212	T414	T800	
ldtimer	2	2	2	
tin	30	30	30	time future
	3	3	3	time past
talt	4	4	4	
taltwt	15	15	15	time past
	48	48	48	time future
enbt	8	8	8	
dist	23	23	23	

Table B.8: Input/output operation codes

Mnemonic	Processor Cycles			Notes
	T212	T414	T800	
in	2w+19	2w+19	2w+19	
out	2w+19	2w+19	2w+19	
outword	23	23	23	
outbyte	23	23	23	
resetch	3	3	3	
alt	2	2	2	
altwt	5	5	5	channel ready
	17	17	17	channel not ready
altend	4	4	4	
enbs	3	3	3	
diss	4	4	4	
enbc	7	7	7	channel ready
	5	5	5	channel not ready
disc	8	8	8	

Table B.9: Control operation codes

Mnemonic	Processor Cycles			Notes
	T212	T414	T800	
ret	5	5	5	
ldpi	2	2	2	
gajw	2	2	2	
dup	n/a	n/a	1	
gcall	4	4	4	
lend	10	10	10	loop
	5	5	5	exit

Table B.10: Scheduling operation codes

Mnemonic	Processor Cycles			Notes
	T212	T414	T800	
startp	12	12	12	
endp	13	13	13	
runp	10	10	10	
stopp	11	11	11	
ldpri	1	1	1	

Table B.11: Error handling operation codes

Mnemonic	Processor Cycles			Notes
	T212	T414	T800	
csub0	2	2	2	
ccnt1	3	3	3	
testerr	2	2	2	no error
	3	3	3	error
seterr	1	1	1	
stoperr	2	2	2	no error
clrhalterr	1	1	1	
sethalterr	1	1	1	
testhalterr	2	2	2	

Table B.12: Processor initialisation operation codes

Mnemonic	Processor Cycles			Notes
	T212	T414	T800	
testpranal	2	2	2	
saveh	4	4	4	
savel	4	4	4	
sthf	1	1	1	
sthb	1	1	1	
stlf	1	1	1	
stlb	1	1	1	
sttimer	1	1	1	

B Instruction set summary

Table B.13: Floating point support operation codes

Mnemonic	Processor Cycles			Notes
	T212	T414	T800	
cflerr	n/a	3	n/a	
unpacksn	n/a	15	n/a	
roundsn	n/a	12/15	n/a	
postnormsn	n/a	5/30	n/a	
ldinf	n/a	1	n/a	

Processor cycles are shown as **Typical/Maximum** cycles.

Table B.14: Floating point load/store operation codes

Mnemonic	Processor Cycles			Notes
	T212	T414	T800	
fpldnlsn	n/a	n/a	2	
fpldnldb	n/a	n/a	3	
fpldnlsni	n/a	n/a	4	
fpldnldbi	n/a	n/a	6	
fpldzerosn	n/a	n/a	2	
fpldzerodb	n/a	n/a	2	
fpldnladdsn	n/a	n/a	8/11	
fpldnladddb	n/a	n/a	9/12	
fpldnlmulsn	n/a	n/a	13/20	
fpldnlmuldb	n/a	n/a	21/30	
fpstnlsn	n/a	n/a	2	
fpstnldb	n/a	n/a	3	
fpstnli32	n/a	n/a	4	

Processor cycles are shown as **Typical/Maximum** cycles.

Table B.15: Floating point general operation codes

Mnemonic	Processor Cycles			Notes
	T212	T414	T800	
fpentry	n/a	n/a	1	
fprev	n/a	n/a	1	
fpdup	n/a	n/a	1	

Table B.16: Floating point rounding operation codes

Mnemonic	Processor Cycles			Notes
	T212	T414	T800	
fpurn	n/a	n/a	1	
fpurz	n/a	n/a	1	
fpurp	n/a	n/a	1	
fpurm	n/a	n/a	1	

Table B.17: Floating point comparison operation codes

Mnemonic	Processor Cycles			Notes
	T212	T414	T800	
fpgt	n/a	n/a	4/6	
fpeq	n/a	n/a	3/5	
fpordered	n/a	n/a	3/4	
fpnan	n/a	n/a	2/3	
fpnotfinite	n/a	n/a	2/2	
fpuchki32	n/a	n/a	3/4	
fpuchki64	n/a	n/a	3/4	

Processor cycles are shown as **Typical/Maximum** cycles.

Table B.18: Floating point conversion operation codes

Mnemonic	Processor Cycles			Notes
	T212	T414	T800	
fpur32tor64	n/a	n/a	3/4	
fpur64tor32	n/a	n/a	6/9	
fprtoi32	n/a	n/a	7/9	
fpi32tor32	n/a	n/a	8/10	
fpi32tor64	n/a	n/a	8/10	
fpb32tor64	n/a	n/a	8/8	
fpunoround	n/a	n/a	2/2	
fpint	n/a	n/a	5/6	

Processor cycles are shown as **Typical/Maximum** cycles.

Table B.19: Floating point arithmetic operation codes

Mnemonic	Processor Cycles						Notes
	T212		T414		T800		
	Single	Double	Single	Double	Single	Double	
fpadd	n/a	n/a	n/a	n/a	6/9	6/9	
fpsub	n/a	n/a	n/a	n/a	6/9	6/9	
fpmul	n/a	n/a	n/a	n/a	11/18	18/27	
fpdiv	n/a	n/a	n/a	n/a	16/28	31/43	
fpuabs	n/a	n/a	n/a	n/a	2/2	2/2	
fpremfirst	n/a	n/a	n/a	n/a	36/46	36/46	
fpremstep	n/a	n/a	n/a	n/a	32/36	32/36	
fpusqrtfirst	n/a	n/a	n/a	n/a	27/29	27/29	
fpusqrtstep	n/a	n/a	n/a	n/a	42/42	42/42	
fpusqrtlast	n/a	n/a	n/a	n/a	8/9	8/9	
fpuexpinc32	n/a	n/a	n/a	n/a	6/9	6/9	
fpuexpdec32	n/a	n/a	n/a	n/a	6/9	6/9	
fpumulby2	n/a	n/a	n/a	n/a	6/9	6/9	
fpudivby2	n/a	n/a	n/a	n/a	6/9	6/9	

Processor cycles are shown as **Typical/Maximum** cycles.

B Instruction set summary

Table B.20: Floating point error operation codes

Mnemonic	Processor Cycles			Notes
	T212	T414	T800	
fpchkerror	n/a	n/a	1	
fptesterror	n/a	n/a	2	
fpuseterror	n/a	n/a	1	
fpuclearerror	n/a	n/a	1	

Table B.21: Block move operation codes

Mnemonic	Processor Cycles			Notes
	T212	T414	T800	
move2dinit	n/a	n/a	8	
move2dall	n/a	n/a	(2**p**+23)∗**r**	
move2dnonzero	n/a	n/a	(2**p**+23)∗**r**	
move2dzero	n/a	n/a	(2**p**+23)∗**r**	

Table B.22: CRC and bit operation codes

Mnemonic	Processor Cycles			Notes
	T212	T414	T800	
crcword	n/a	n/a	35	
crcbyte	n/a	n/a	11	
bitcnt	n/a	n/a	**b**+2	
bitrevword	n/a	n/a	36	
bitrevnbits	n/a	n/a	**n**+4	

Appendix C

bibliography

C bibliography

C   Bibliography

This appendix contains a list of some transputer-related publications and journal or magazine articles which may be of interest to the reader. The References section details publications referred to in this manual, other than the standard INMOS documents detailed below. The INMOS Authors and Other Authors sections contain only a small sample of published articles concerning the Transputer and occam during the past two years.

INMOS publish manuals and data sheets pertaining to transputer based products and to occam. Apart from items detailed below, INMOS produce an Engineering Data Sheet for each product, as well as Product Information Guides detailing the INMOS range of products. There are also a number of Technical Notes and Application Notes available from INMOS, covering a wide range of topics in both the hardware and software areas.

C.1   INMOS publications

INMOS
   occam *Programming Manual*
   72 OCC 040

INMOS
   occam *2 Language Definition*
   72 OCC 044

INMOS
   Transputer Development System Manual
   72 TDS 141

INMOS
   IMS M212 Product Data manual
   72 TRN 103

INMOS
   The Transputer Instruction Set - A Compiler Writers' Guide
   72 TRN 119

INMOS
   occam *2 Reference Manual*
   Prentice Hall
   ISBN 0-13-629312-3

INMOS
   occam
   Keigaku Shuppan Publishing Company
   ISBN 4-7665-0133-0
   (In Japanese)

C.2 INMOS technical notes

A wide range of technical and application notes are published by INMOS. Some of these are listed in this section.

INMOS
> *IMS T800 Architecture*
> Technical note 6
> 72 TCH 006

D May
> *Communicating processes and* occam
> Technical note 20
> 72 TCH 020

D May and R Shepherd
> *The transputer implementation of* occam
> Technical note 21
> 72 TCH 021

D May and R Shepherd
> *Communicating Process Computers*
> Technical note22
> 72 TCH 022

D May and C Keane
> *Compiling* occam *into silicon*
> Technical note23
> 72 TCH 023

N Miller
> *Exploring Multiple Transputer Arrays*
> Technical note24
> 72 TCH 024

S Ericsson Zenith
> occam *2: aspects of the language and its development*
> Technical note25
> 72 TCH 025

G Harriman
> *Notes on Graphics Support and Performance Improvements on the IMS T800*
> Technical note26
> 72 TCH 026

C.3 Papers and extracts by INMOS authors

I M Barron
>   *The transputer and* occam
>   Information Processing 86 (IFIP congress)
>   (Dublin, Ireland, 1-5 September 1986)
>   pp 259-265

M Homewood, D May, D Shepherd, R Shepherd
>   *The IMS T800 transputer*
>   IEEE Micro
>   Vol 7 No 5, pp 10-26, October 1987

D May, R Shepherd and C Keane
>   *Communicating Process Architecture: Transputers and* occam
>   Lecture Notes in Computer Science: Future Parallel Computers
>   (Pisa Summer School June 1986)
>   Springer-Verlag
>   272 pp 35-81 June 1986
>   ISBN 3-540-18203-9, ISBN 0-387-18203-9
>   (Also published as INMOS Technical Notes 20, 21, 22)

D May and D Shepherd
>   *Formal verification of the IMS T800 microprocessor*
>   Electronic Design Automation Conference
>   (Wembley, 13-16 July 1987)
>   pp 605-615

D May and R Shepherd
>   *The INMOS transputer*
>   Parallel processing: state of the art report
>   Pergamon Infotech Ltd
>   ISBN 0-08-034113-6

P Walker
>   *The transputer: a building block for parallel processing*
>   Byte
>   Vol 10 No 5, pp 219-235, May 1985

C.4 Papers and extracts by other authors

H Aiso
Parallelism in new generation computing
ICOT Journal
7, pp 12-35, March 1985

C R Askew et al
Simulation of statistical mechanical systems on transputer arrays
Computer Physics Communications
42, pp 21-26, 1986

M J P Bolton and D A Cowling
Real-time flight simulation with transputers
Modeling and Simulation. Sixteen Annual Pittsburgh Conference
(Pittsburgh, PA, USA. 25-26 April 1985)
pp 911-916

D S Broomhead et al
A practical comparison of the systolic and wavefront array processing architectures
ICASSP 85, IEEE International Conference on Acoustics Speach and Signal Processing
(Tampa, Florida, 26-29 March 1985)
1, pp 296-9

R A Browse and D B Skillicorn
An implementation of the parallelism in visual object interpretation
PCCC86, Fifth Annual International Pheonix Conference on Computers and Communications
(Scotsdale, AZ, 26-28 March 1986)
pp 487-490

R S Cok
Case study: multiprocessing with transputers
Electronic Engineering Times
373, T15-T16, 17 March 1986

P Eckelmann
Transputer der 2. Generation (1. Teil)
Elektronik
H. 18, S. 61-70 1987

P Eckelmann
Transputer der 2. Generation (2. Teil)
Elektronik
H. 19, S. 129-136 1987

A J Fisher
A multi-processor implementation of occam
Software - Practice and Experience
Vol 16 No 10, pp 875-892, October 1986

B M Forrest et al
Implementing neural network models on parallel computers
The Computer Journal
Vol 30 No 5, pp 413-419, October 1987

J G Harp et al
Signal processing with transputer arrays (TRAPS)
Computer Physics Communications
37(1-3), pp 77-86, July 1985

C Bibliography

A J G Hey and J S Ward
  Design of a high performance multiprocesor machine based on transputers with applications to Monte Carlo simulations
  Paris Conf on Advances on Reactor Physics, Mathematics and Computation 1987

A J G Hey et al
  High performance simulation of lattice physics using transputer arrays
  Computing in High Energy Physics
  North Holland, 1986
  ISBN 0-444-87973-0

S Y Kung
  On programming languages for VLSI array processors
  Conference on Highly Parallel Signal Processing Architectures
  (Los Angeles, CA, 21-22 January 1986)
  Proceedings of SPIE - International Society of Optical Engineering
  614, pp 118-133, 1986

K Leppala
  Utilization of parallelism in transputer-based real-time control systems
  Microprocessing and Microprogramming
  21(1-5), pp 629-636, 1987

T Mano et al
  occam *to CMOS: experimental logic design support system*
  IFIP 7th International Symposium on Computer Hardware Description Languages
  (Tokyo, 29-31 August 1985)
  pp 301-390

T Manuel and S Rogerson
  The transputer finally starts living up to its claims
  Electronics
  60(17), pp 78-80, 20 August 1987

T Manuel and S Rogerson
  INMOS puts transputers into its own CAD system
  Electronics
  60(17), pp 81-82, 20 August 1987

R M Marshall
  Automatic generation of controller systems from control software
  ICCAD '86, IEEE International Conference on Computer-Aided Design
  (Santa Clara, CA, 11-13 November 1986)
  pp 256-9

D L McBurney and M R Sleep
  Transputer-based experiments with ZAPP architecture
  Lecture Notes in Computer Science: PARLE. Parallel Architectures and Languages
  (Eindhoven June 1987)
  Springer-Verlag
  258 Vol 1 pp 242-259 June 1987
  ISBN 3-540-17943-7, ISBN 0-387-17943-7

G McIntire et al
  Design of a neural network simulator on a transputer array
  Space Operations - Automation and Robotics Workshop 87
  (NASA/Johnson Space Centre, Houston, TX, 5-7 August 1987)

J R Newport
> *The INMOS transputer*
> 32-bit microprocessors (edited by H J Mitchell)
> Collins, 1986
> pp 93-129 ISBN 0-00-383067-5

D Pountain
> *Turbocharging Mandlebrot*
> Byte
> Vol 11 No 9, pp 359-366, September 1986

D J Pritchard, C R Askew, D B Carpenter, I Glendinning, A J G Hey, D A Nicole
> *Practical parallelism using transputer arrays*
> Lecture Notes in Computer Science: PARLE. Parallel Architectures and Languages
> (Eindhoven June 1987)
> Springer-Verlag
> 258 Vol 1 pp 278-294 June 1987
> ISBN 3-540-17943-7, ISBN 0-387-17943-7

M Schindler
> *An der Schwelle zur 2. Computer-Ara*
> Elektronik
> H. 10, S. 73-80 1987

J Stender
> *Parallele prolog implementierung auf transputer*
> Hard and Soft
> pp 20-23, September 1987
> (In German)

K Uedu
> *Parallel programming languages*
> Information Processing Society of Japan (Joho Shori)
> 27(9), pp 995-1004, 1986
> (In Japanese)

P H Welch
> *Emulating digital logic using transputer networks*
> Lecture Notes in Computer Science: PARLE. Parallel Architectures and Languages
> (Eindhoven June 1987)
> Springer-Verlag
> 258 Vol 1 pp 357-373 June 1987
> ISBN 3-540-17943-7, ISBN 0-387-17943-7

D Wilson (editor)
> *Transputer spawns a new class of applications*
> Digital Design
> 16(12), pp 34-44, November 1986

C.5 Books and monographs

G Barrett
: *Formal methods applied to a floating point number system*
Oxford University Computing Laboratory Programming Research Group
Technical Monograph PRG-58

INMOS
: **occam** *2 reference manual*
Prentice Hall
ISBN 0-13-629312-3

INMOS
: **occam**
Keigaku Shuppan Publishing Company
ISBN 4-7665-0133-0
(In Japanese)

C Jesshope, R J O'Gorman, J M Stewart (editors)
: *Parallel processing: state of the art report*
Pergamon Infotech Ltd
ISBN 0-08-034113-6

G Jones
: *Programming in* **occam**
Prentice Hall
ISBN 13-729773-4

J Kerridge
: **occam** *programming: a practical approach*
Blackwell Scientific Publications
ISBN 0-632-01659-0

Onai
: **occam** *and transputer*
Kyoshin Publishing Co.
ISBN 4-320-02269-6
(In Japanese)

D Pountain and D May
: *A tutorial introduction to* **occam** *programming*
Blackwell Scientific Publications/McGraw-Hill
ISBN 0-632-01847-X, ISBN 0-07-050606-X

D Pountain and R Rudolph
: **occam**, *das Handbuch*
Verlag Heinz Heise GmbH
ISBN 3-88229-001-3
(In German)

A W Roscoe and C A R Hoare
: *The laws of* **occam** *programming*
Oxford University Computing Laboratory Programming Research Group
Technical Monograph PRG-53

A W Roscoe and N Dathi
: *The pursuit of deadlock freedom*
Oxford University Computing Laboratory Programming Research Group
Technical Monograph PRG-57

T A Theoharsis
> *Exploiting parallelism in the graphics pipeline*
> Oxford University Computing Laboratory Programming Research Group
> Technical Monograph PRG-54

C.6 References

This section details publications referred to in the TRANSPUTER OVERVIEW chapter of this document.

Harp
> *Phase 1 of the development and application of a low cost, high performance multiprocessor machine*
> J G Harp et al
> ESPRIT '86: Results and Achievements, Elsevier Science Publishers B.V.
> pp 551-562

IEEE
> *IEEE Standard for Binary Floating-Point Arithmetic*
> ANSI/IEEE Std 754-1985.

INMOS '84
> *IMS T414 reference manual*
> INMOS Limited 1984

INMOS '87
> *The Transputer Instruction Set - A Compiler Writers' Guide*
> INMOS Ltd 72 TRN 119

McMahon
> *The Livermore Fortran Kernels: A Computer Test of the Numerical Performance Range*
> F H McMahon
> Lawrence Livermore National Laboratory
> UCRL-53745

Appendix D

index

Appendix D

index

D   Index

! 9
" 15
() 14
* 14, 298, 301
\+ 14, 298, 301
− 14, 298, 301
/ 14, 298, 301
/\ 14, 298
:= 9
< 14, 298, 301
<< 14, 298
<= 14, 298, 301
<> 14, 298, 301
= 14, 298, 301
\> 14, 298, 301
>< 14, 298
>= 14, 298, 301
>> 14, 298
? 9
\/ 14, 298
~ 14

ABS 302
Absolute maximum ratings
  IMS C004   242
  IMS C011   266
  IMS C012   288
  IMS T212   204
  IMS T414   156
  IMS T800   99
Access
  byte-wide IMS C011   251
  byte-wide IMS C012   275
Acknowledge
  link   **36**
  link IMS C004   234
  link IMS C011   256
  link IMS C012   280
  link IMS T212   201
  link IMS T414   153
  link IMS T800   96
Address
  bus IMS T212   **190**
  bus IMS T414   **134**
  bus IMS T800   **77**
  byte IMS T212   187
  byte IMS T414   130
  byte IMS T800   73
  mark IMS M212   **215**
  refresh IMS T414   147
  refresh IMS T800   90
  space IMS T212   167, 169
  space IMS T414   109, 111
  space IMS T800   47, 49
AFTER   14, **15**, 37, 299

ALT   **11**, 12, 18, 26, 299, 303
Alternation construction   10, **11**, 12, 26, 37
Analyse
  IMS T212   **185**, 186
  IMS T414   **128**, 129
  IMS T800   **71**, 72
AND   **14**, 298
ANSI-IEEE 754-1985   13, 39
  IMS T800   47, 65
Application
  bidirectional exchange IMS C004   238
  bus systems IMS C004   238
  drawing coloured text   41
  enhanced controller IMS M212   220
  IMS T212   193
  IMS T414   140
  IMS T800   83
  link switching IMS C004   238
  multiple control IMS C004   238
  winchester controller IMS M212   219, 220
Architecture   **25**, 26
  internal   **28**
  rationale   **5**
Arithmetic
  multiple length   **300**
  operation IMS T212   169
  operation IMS T414   111
  operation IMS T800   49
  operator   298, 301
Array   **13**
  assignment   303
  byte   15
  of disk controllers IMS M212   **221**
  of processes   **12**
  of transputers IMS M212   221
  type   **13**
  variable   298
ASCII   15
ASHIFTLEFT   300
ASHIFTRIGHT   300
Assignment   9, 26, 298, 301
  array   303
  process   8, **9**

Bandwidth
  memory   41
  memory IMS T212   171
  memory IMS T414   113
  memory IMS T800   51
Barrel shifter   39
Behaviour
  logical   **3**, 16, **18**
  physical   **3**
Benchmark
  LINPACK   39

speed 38
Whetstone 38, 39
Bit
  counting performance 302
  data **36**
  operator 298
  reversal performance 302
  start **36**
  stop **36**
Bit-blt 41
BITCOUNT 302
BITREVNBIT 302
BITREVWORD 302
Block move **41**
  conditional 41
  IMS T800 47
  performance 302
  two-dimensional 41
BOOL 13
Boolean
  expression 303
  operator 298
BootFromRom
  IMS T212 **183**, 185
  IMS T414 **126**, 128
  IMS T800 **69**, 71
Bootstrap 20, 21
  address IMS T212 185
  address IMS T414 128
  address IMS T800 71
  code IMS T212 **187**
  code IMS T414 **130**
  code IMS T800 **73**
  IMS M212 **217**
  IMS T212 **183**, **185**
  IMS T414 **126**, 128, 150
  IMS T800 **69**, **71**, 93
  program IMS T212 199
Brackets **14**
Buffer
  input IMS C011 264
  input IMS C012 286
  link 20
  output IMS C011 264
  output IMS C012 286
Bus 28
  IMS C011 261
  IMS C012 283
Byte
  access IMS C011 251
  access IMS C012 275
  access IMS T212 190, **195**
  address IMS T212 187, 190
  address IMS T414 130
  address IMS T800 73
BYTE 13

C 19
Capacitive load 5

CapMinus
  IMS C004 **231**
  IMS C011 **253**
  IMS C012 **277**
  IMS T212 **182**
  IMS T414 **124**
  IMS T800 **67**
CapPlus
  IMS C004 **231**
  IMS T212 **182**
  IMS T414 **124**
  IMS T800 **67**
CASE 12
CHAN OF 13
  protocol 298
Channel 4, 8, 9, 11, 13, 16, 19, 26, 298
  communication 33, 35
  disk hardware IMS M212 216
  empty 33
  event IMS T212 **200**
  event IMS T414 **152**
  event IMS T800 **95**
  external 33
  external IMS T212 174
  external IMS T414 116
  external IMS T800 54
  IMS T212 174
  IMS T414 116
  IMS T800 54
  input 26
  internal 33
  internal IMS T212 174
  internal IMS T414 116
  internal IMS T800 54
  link 19
  link IMS C004 234
  link IMS C011 256
  link IMS C012 280
  link IMS T212 201
  link IMS T414 153
  link IMS T800 96
  memory 19
  occam 19
  output 26
  process 26
Characteristics
  AC timing IMS C004 244
  AC timing IMS C011 268
  AC timing IMS C012 290
  AC timing IMS T212 206
  AC timing IMS T414 158
  AC timing IMS T800 101
  DC electrical IMS C004 242, 243
  DC electrical IMS C011 266, 267
  DC electrical IMS C012 288, 289
  DC electrical IMS T212 204, 205
  DC electrical IMS T414 156, 157
  DC electrical IMS T800 99, 100
CLIP2D 41, 302
Clock 13, 21

input 20, 21
  input, internal IMS C004 **231**
  input, internal IMS C011 **253**
  input, internal IMS C012 **277**
  input, internal IMS T212 **182**
  input, internal IMS T414 **124**
  input, internal IMS T800 **67**
  internal 20
  link 21
  link IMS C004 234
  link IMS C011 256
  link IMS C012 280
  link IMS T212 201
  link IMS T414 153
  link IMS T800 96
  multiple IMS C004 231
  multiple IMS C011 253
  multiple IMS C012 277
  multiple IMS T212 182
  multiple IMS T414 124
  multiple IMS T800 67
  phase 7
  processor 37
  processor IMS T212 **174**
  processor IMS T414 **116**
  processor IMS T800 **54**
  stability IMS C004 231
  stability IMS C011 253
  stability IMS C012 277
  stability IMS T212 182
  stability IMS T414 124
  stability IMS T800 67
  timer 15
  timer IMS T212 **174**
  timer IMS T414 **116**
  timer IMS T800 **54**
  transputer 7
ClockIn
  IMS C004 **231**
  IMS C011 **253**
  IMS C012 **277**
  IMS T212 **182**
  IMS T414 **124**
  IMS T800 **67**
  period IMS T212 **189**
  period IMS T414 **132**
  period IMS T800 **75**
  skew IMS C011 254
Code
  function/operation IMS T212 **176**
  function/operation IMS T414 **118**
  function/operation IMS T800 **56**
Coding efficiency
  IMS T212 171
  IMS T414 113
  IMS T800 51
Colour
  display 41
  graphics 31
  text example 41

Communication 4, 5, **6**, 25, 28, **33**
  bandwidth **5**
  channel 8, 33, 35
  construction 8, **11**
  contention 5
  external 35
  frequency 7, 20
  IMS T212 173
  IMS T414 115
  IMS T800 53
  interface 6
  internal 33
  language 19
  link 5, **36**
  parallel IMS C011 251
  parallel IMS C012 275
  process IMS T212 **174**
  process IMS T414 **116**
  process IMS T800 **54**
  speed 27
Comparison operator 298, 301
Compatibility
  IMS T800 **47**
Concept 26
Concurrency 3, 8, 25
  IMS T212 **172**
  IMS T414 **114**
  IMS T800 **52**
  internal 8
  support **31**
Concurrent
  FPU/CPU operation 38
  process 8, 10, 11, 18, 26
  systems 8
Conditional construction 10, **11**
Configuration
  coding IMS T414 146
  coding IMS T800 89
  memory IMS T212 185
  memory IMS T414 128, 132, **141**
  memory IMS T800 71, 75, **84**
  memory, external IMS T414 141, **142**, 143, 144, 145, 147
  memory, external IMS T800 84, **85**, 86, 87, 88, 90
  memory, internal IMS T414 **141**
  memory, internal IMS T800 **84**
  program **16**
  refresh coding IMS T414 146
  refresh coding IMS T800 89
Connection
  link IMS C004 234
  link IMS C011 256
  link IMS C012 280
  link IMS T212 201
  link IMS T414 153
  link IMS T800 96
Constant 299
  subscript 298
  value **30**

Construction 10, 26, 299
  alternation 10, **11**, 12, 26, 37
  communication **11**
  conditional 10, **11**
  parallel 8, **10**, **16**, 26, 32
  parallel IMS T212 **173**
  parallel IMS T414 **115**
  parallel IMS T800 **53**
  performance 303
  repetition **12**
  replication **12**
  selection **12**
  sequential 8, **10**, 11, 26
Control
  byte IMS T212 **185**
  byte IMS T414 126, **128**
  byte IMS T800 **71**
  link IMS C004 233
  logic IMS M212 216
Conversion
  `INT, REAL` 301
  `REAL, INT` 301
CPU **28**
  concurrent operation 38
  register **28**, **29**
CRC
  IMS M212 **215**, **217**
  IMS T800 47
  performance 302
CRCBYTE 302
CRCWORD 302
Cyclic redundancy
  IMS M212 **215**, **217**
  IMS T800 47
  performance 302

D0-7
  IMS C011 255, **261**
  IMS C012 279, **283**
DABS 302
Data
  bit **36**
  bus IMS T212 189, **190**
  bus IMS T414 **134**
  bus IMS T800 **77**
  link **36**
  link IMS C004 234
  link IMS C011 256
  link IMS C012 280
  link IMS T212 201
  link IMS T414 153
  link IMS T800 96
  rate 20, 36
  rate IMS T212 167
  rate IMS T414 109
  rate IMS T800 47
  rate link IMS T212 201
  rate link IMS T414 153
  rate link IMS T800 96
  read IMS C011 **265**

read IMS C012 **287**
separation IMS M212 **216**
serial **36**
structure 30
structure IMS T212 170
structure IMS T414 112
structure IMS T800 50
transfer 11
value 29
value IMS T212 170
value IMS T414 112
value IMS T800 50
write IMS C011 **265**
write IMS C012 **287**
Data Present
  IMS C011 261, 264, 265
  IMS C012 283, 286, 287
Declaration **13**, 298
Decoupling
  IMS C004 **231**
  IMS C011 **253**
  IMS C012 **277**
  IMS T212 **182**
  IMS T414 **124**
  IMS T800 **67**
Delay
  input 15
  timer 15
Deschedule **32**, 34, 35
  IMS T212 **172**, 173, 174
  IMS T414 **114**, 115, 116
  IMS T800 **52**, 53, 54
  point IMS T212 **173**, **177**, 185
  point IMS T414 **115**, **119**, 128
  point IMS T800 **53**, **57**, 71
Device 26
Direct function **30**
  IMS T212 **170**
  IMS T414 **112**
  IMS T800 **50**
Direct memory access
  IMS T212 190
  IMS T414 133
  IMS T800 76
Disk
  command IMS M212 217, 218
  compression/decompression IMS M212 221
  controller IMS M212 215
  cylinder IMS M212 217
  drive selection IMS M212 215
  encryption/decryption IMS M212 221
  floppy IMS M212 215, **217**
  format IMS M212 **217**
  head IMS M212 217
  head position IMS M212 215
  interleave IMS M212 218
  management IMS M212 221
  parameter IMS M212 217
  port IMS M212 215
  programming interface IMS M212 **217**

SA400/450 IMS M212   215, **217**
sector IMS M212   217
ST506/412 IMS M212   215, **217**
status IMS M212   215
winchester IMS M212   215, **217**
DMA
  at reset IMS T212   **198**, 199
  at reset IMS T414   **151**
  at reset IMS T800   **94**
  IMS T212   190, 191, **198**
  IMS T414   133, **150**
  IMS T800   76, **93**
  operation IMS T212   **198**
  operation IMS T414   **151**
  operation IMS T800   **94**
DRAW2D   41, 302
DSQRT   302

ECC
  IMS M212   **215, 217**
Efficiency   **31**
Electrical
  AC timing characteristics IMS C004   244
  AC timing characteristics IMS C011   268
  AC timing characteristics IMS C012   290
  AC timing characteristics IMS T212   206
  AC timing characteristics IMS T414   158
  AC timing characteristics IMS T800   101
  DC characteristics IMS C004   242, 243
  DC characteristics IMS C011   266, 267
  DC characteristics IMS C012   288, 289
  DC characteristics IMS T212   204, 205
  DC characteristics IMS T414   156, 157
  DC characteristics IMS T800   99, 100
  operating conditions IMS C004   242
  operating conditions IMS C011   266
  operating conditions IMS C012   288
  operating conditions IMS T212   204
  operating conditions IMS T414   156
  operating conditions IMS T800   99
  specification   20
EMI
  IMS T212   **189**
  IMS T414   **132**
  IMS T800   **75**
Equivalent circuit
  IMS C004   243
  IMS C011   267
  IMS C012   289
  IMS T212   205
  IMS T414   157
  IMS T800   100
Erastosthenes   303
Error   21
  IMS T212   **186**
  IMS T414   **129**
  IMS T800   **72**
Error   21
  IMS T212   **186**
  IMS T414   **129**

IMS T800   **72**
power up IMS T212   186
power up IMS T414   129
power up IMS T800   72
Error   21
  analysis   17
  analysis IMS T212   186
  analysis IMS T414   129
  analysis IMS T800   72
  circuit IMS T212   186
  circuit IMS T414   129
  circuit IMS T800   72
  correcting code   31
  correcting code IMS M212   **215, 217**
  expression check   299
  floating point IMS T800   **66**
  handling   17
  IMS M212   217
  languages   17
  reset IMS T212   186
  reset IMS T414   129
  reset IMS T800   72
ErrorIn
  IMS T800   **72**
Evaluation
  expression IMS T212   **169**, 171
  expression IMS T414   **111**, 113
  expression IMS T800   **49**, 51
  stack   28, **29**, 33
  stack IMS T212   **169**, 173, 174
  stack IMS T414   **111**, 115, 116
  stack IMS T800   **49**, 53, 54
Event   12, 22
  IMS T212   **200**
  IMS T414   **152**
  IMS T800   **95**
EventAck
  IMS T212   **200**
  IMS T414   **152**
  IMS T800   **95**
EventReq
  IMS T212   186, **200**
  IMS T414   129, **152**
  IMS T800   72, **95**
Example
  drawing coloured text   41
  instruction set IMS T212   176
  instruction set IMS T414   118
  instruction set IMS T800   56
Execution
  instruction IMS T212   170
  instruction IMS T414   112
  instruction IMS T800   50
Expression   8, **14**, 26, 298, 299, 301
  evaluation IMS T212   **169**, 171
  evaluation IMS T414   **111**, 113
  evaluation IMS T800   **49**, 51
  subscript   298
External
  memory interface IMS T212   **189**

memory interface IMS T414 **132**
memory interface IMS T800 **75**
memory performance **303**
registers 15

Factorial 14
FALSE **14**
Flash multiplier 39
Floating point 25, **37**
 address 37
 arithmetic 39
 co-processor 39
 comparison 39
 concurrency IMS T800 **65**
 concurrent operation 38
 datapath 39
 design 37, **39**
 division 39
 double length IMS T800 **65**
 error IMS T800 **66**
 functions **302**
 instruction 31, **37**
 microcode 39
 multiplication 38, 39
 normalise IMS T800 **65**
 operand 37
 performance 39, 297, **300**, **301**
 processor **28**, **37**
 processor IMS T800 47, **65**
 register **28**
 rounding IMS T800 **65**
 selector sequence IMS T800 **56**, **65**
 single length IMS T800 **65**
 speed 39
 stack IMS T800 **65**
 standard 39
Floating point numbers 13
FM
 IMS M212 **215**
FOR **12**
Fortran 19
FPU (see Floating point) 28, 65
FP'Error
 IMS T800 **66**
FRACMUL **300**
Frequency
 changer IMS C011 251
 ClockIn IMS C004 231
 ClockIn IMS C011 253
 ClockIn IMS C012 277
 ClockIn IMS T212 182
 ClockIn IMS T414 124
 ClockIn IMS T800 67
 link 20
 modulation IMS M212 **215**
Function **14**, 299
 code **29**
 code IMS T212 **170**, 176
 code IMS T414 **112**, 118
 code IMS T800 **50**, 56

direct 30
direct IMS T212 **170**
direct IMS T414 **112**
direct IMS T800 **50**
indirect 31
indirect IMS T212 **171**
indirect IMS T414 **113**
indirect IMS T800 **51**
prefix **30**, 31
prefix IMS T212 **170**
prefix IMS T414 **112**
prefix IMS T800 **50**
FUNCTION **14**, 298, 301

GND 20
 IMS C004 **231**
 IMS C011 **253**
 IMS C012 **277**
 IMS T212 **182**
 IMS T414 **124**
 IMS T800 **67**
Graphics **41**
 support IMS T800 **47**

Halt 17
 IMS T212 185, 186
 IMS T414 128, 129
 IMS T800 71, 72
HaltOnError
 IMS T212 **186**
 IMS T414 **129**
 IMS T800 **72**
Handshake 8
 event IMS T212 **200**
 event IMS T414 **152**
 event IMS T800 **95**
 parallel IMS C011 251, 259, 260
Hardware **5**
 channel IMS M212 216
 IMS M212 215, 218
Harness 8, 19

I0-7
 IMS C011 259
IAck
 IMS C011 255, 259
IF **11**, 12, 26, 299, 303
Implementation
 hard-wired **5**
 hardware **5**
 link **6**
 occam **5**
 program 8
IMS B005 220
IMS C004 **229**
IMS C011 **251**
IMS C012 **275**
IMS M212 **215**
IMS T212 **167**
IMS T414 39, 41, **109**

link 36
IMS T800 38, 39, 41, **47**
  block move 41
  floating point 37, 39
  link 36
Indirect function **31**
  IMS T212 **171**
  IMS T414 **113**
  IMS T800 **51**
Indirection code
  instruction IMS T800 **65**
Input 8, 9, 15, 21, 26, 298, 301
  buffer IMS C011 264
  buffer IMS C012 286
  channel 26
  clock 20, 21
  clock IMS C004 **231**
  clock IMS C011 **253**
  clock IMS C012 **277**
  clock IMS T212 **182**
  clock IMS T414 **124**
  clock IMS T800 **67**
  link IMS C004 **234**
  link IMS C011 **256**
  link IMS C012 **280**
  link IMS T212 185, **201**
  link IMS T414 128, **153**
  link IMS T800 71, **96**
  pins 20
  port IMS C011 **259**
  process 8, **9**, 11
  process IMS T212 174
  process IMS T414 116
  process IMS T800 54
  register IMS C011 **261**, 264
  register IMS C012 **283**, 286
  timer 299
  voltage 20
InputInt
  IMS C011 255, **264**, 265
  IMS C012 279, **286**, 287
Instruction
  arithmetic IMS T212 171
  arithmetic IMS T414 113
  arithmetic IMS T800 51
  comparison IMS T212 171
  comparison IMS T414 113
  comparison IMS T800 51
  descheduling IMS T212 **177**
  descheduling IMS T414 **119**
  descheduling IMS T800 **57**
  description 29, 30, 31, 32, 33, 37
  description IMS T212 169, **170**, 171, 173, 174
  description IMS T414 111, **112**, 113, 115, 116
  description IMS T800 49, **50**, 51, 53, 54, 56, 65, 66
  error IMS T212 **177**
  error IMS T414 **119**
  error IMS T800 **58**
  execution IMS T212 170
  execution IMS T414 112
  execution IMS T800 50
  floating point **37**
  floating point error IMS T800 **58**
  format IMS T212 **170**
  format IMS T414 **112**
  format IMS T800 **50**
  IMS T212 169
  IMS T414 111
  IMS T800 49
  indirection code IMS T800 **65**
  logical IMS T212 171
  logical IMS T414 113
  logical IMS T800 51
  memory relative IMS T212 170
  memory relative IMS T414 112
  memory relative IMS T800 50
  operation **30**
  pointer 32
  pointer IMS T212 173
  pointer IMS T414 115
  pointer IMS T800 53
  prefetch 31
  single byte IMS T212 **170**
  single byte IMS T414 **112**
  single byte IMS T800 **50**
  workspace IMS T212 173
  workspace IMS T414 115
  workspace IMS T800 53
Instruction set 8, **29**, **178**
  comparison 307
  design **29**
  example IMS T212 **176**
  example IMS T414 **118**
  example IMS T800 **56**
  IMS T212 169, 170, 171, 176
  IMS T414 111, 112, 113, 118, **120**
  IMS T800 49, 50, 51, 56, **59**
`INT` **13**, 298
`INT16` **13**
`INT32` **13**
  conversion 301
`INT64` **13**
  conversion 301
Integer performance 297
Integrated memory 26
Interface
  application specific 21
  communication 6
  disk controller IMS M212 218
  disk programming IMS M212 **217**
  link 7, **36**
  link IMS C004 **234**
  link IMS C011 **256**
  link IMS C012 **280**
  link IMS T212 **201**
  link IMS T414 **153**
  link IMS T800 **96**
  memory 6, 20
  memory IMS T212 169, 185, **189**

memory IMS T414   111, 128, **132**
memory IMS T800   49, 71, **75**
parallel IMS C011   **261**
parallel IMS C012   **283**
peripheral IMS M212   215
SCSI IMS M212   **222**
serial data IMS M212   215
Interrupt   8, 12, 28
  IMS C011   251, 255
  IMS C012   275
  latency IMS T212   **173**
  latency IMS T414   **115**
  latency IMS T800   **53**
  latency performance   **304**
Interrupt Enable
  IMS C011   264, 265
  IMS C012   286, 287
IntSaveLoc
  IMS T212   **187**
  IMS T414   **130**
  IMS T800   **73**
IS   **14**
IValid
  IMS C011   259

Language   3, 8, 18, 19
  communication   19
  error   17
  IMS T212   170, 171
  IMS T414   112, 113
  IMS T800   50, 51
Latency   **304**
  interrupt IMS T212   **173**
  interrupt IMS T414   **115**
  interrupt IMS T800   **53**
  process IMS T212   **173**
  process IMS T414   **115**
  process IMS T800   **53**
Link   6, 20, 21
  acknowledge   6, **36**
  acknowledge IMS C011   260, 265
  acknowledge IMS C012   287
  acknowlege overlap   36
  adaptor   7, 21, 27
  adaptor IMS C011   251
  adaptor IMS C012   275
  bootstrap ID IMS T212   185
  bootstrap ID IMS T414   128
  bootstrap ID IMS T800   71
  bootstrap IMS T212   183, 185
  bootstrap IMS T414   126
  bootstrap IMS T800   69, 71
  buffer   20
  buffer delays   20
  buffer IMS T212   185
  buffer IMS T414   128
  buffer IMS T800   71
  channel   19
  clock   21
  communication   5, **36**

control IMS C004   233
crossbar switch IMS C004   **229**
data   6, **36**
data IMS C011   260, 261, 264, 265
data IMS C012   283, 286, 287
disk IMS M212   216
frequency   20
implementation   **6**
IMS C004   **234**
IMS C011   251, **256**
IMS C012   275, **280**
IMS T212   174, 185, **201**
IMS T414   116, 128, **153**
IMS T800   54, 71, **96**
input IMS C011   265
input IMS C012   287
input IMS T212   185
input IMS T414   128
input IMS T800   71
interface   35, **36**
interface register   35
message   6
Mode 1 IMS C011   254, **259**
Mode 2 IMS C011   254, **261**
mode select IMS C011   **254**
output IMS T212   185
output IMS T414   128
output IMS T800   71
packet   **36**
parallel adaptor IMS C011   251
parallel adaptor IMS C012   275
peek IMS T212   185
peek IMS T414   128
peek IMS T800   71
performance   303
poke IMS T212   185
poke IMS T414   128
poke IMS T800   71
programmable switch IMS C004   **229**
protocol   6, 7, **36**
signal   20
speed   36
speed IMS C011   251
speed IMS C012   275
speed select IMS C011   **254**
standard   16, 20
start bit   **6**
static IMS T212   170
static IMS T414   112
static IMS T800   50
stop bit   **6**
transfer IMS T212   185
transfer IMS T414   128
transfer IMS T800   71
transmission   36
transputer   8
wiring   27
Link switch
  bit time delay IMS C004   237
  configuration IMS C004   229, **237**

configuration message IMS C004 **237**, 247
implementation IMS C004 **237**
multiplexors IMS C004 **237**
Linked list 32
  IMS T212 **172**
  IMS T414 **114**
  IMS T800 **52**
LinkIn
  IMS C004 234
  IMS C011 255, **256**
  IMS C012 279, **280**
  IMS T212 **201**
  IMS T414 **153**
  IMS T800 **96**
LinkOut
  IMS C004 234
  IMS C011 255, **256**
  IMS C012 279, **280**
  IMS T212 **201**
  IMS T414 **153**
  IMS T800 **96**
LinkSpecial
  IMS T212 **201**
  IMS T414 **153**
  IMS T800 **96**
LINPACK benchmark 39
List
  linked IMS T212 **172**
  linked IMS T414 **114**
  linked IMS T800 **52**
  process IMS T212 **172**, 173
  process IMS T414 **114**, 115
  process IMS T800 **52**, 53
Literal value **30**
  IMS T212 170
  IMS T414 112
  IMS T800 50
Livermore loop 38
Load
  capacitive 5
  instruction IMS T212 170
  instruction IMS T414 112
  instruction IMS T800 50
Logical
  address IMS M212 217
  behaviour **18**
  operation IMS T212 169
  operation IMS T414 111
  operation IMS T800 49
LONGADD 300
LONGDIFF 300
LONGDIV 300
LONGPROD 300
LONGSUB 300
LONGSUM 300
Loop **12**

Map
  memory 19, 21
  process 8

MemA0-15
  IMS T212 **190**, 198
MemAD2-31
  IMS T414 133, **134**, 141, 147
  IMS T800 76, **77**, 84, 90
MemBAcc
  IMS T212 **195**
MemConfig
  IMS T414 **141**, 142
  IMS T800 **84**, 85
MemD0-15
  IMS T212 **190**, 195, 198
MemGranted
  IMS T212 **198**
  IMS T414 **150**
  IMS T800 **93**
MemnotRfD1
  IMS T414 133, **134**, 141, 147
  IMS T800 76, **77**, 84, 90
MemnotWrD0
  IMS T414 133, **134**, 141, 142, 147
  IMS T800 76, **77**, 84, 85, 90
Memory 28
  access IMS T212 169
  access IMS T414 111
  access IMS T800 49
  address IMS T414 **134**
  address IMS T800 **77**
  bandwidth 6, 41
  bandwidth IMS T212 171
  bandwidth IMS T414 113
  bandwidth IMS T800 51
  channel 19
  configuration IMS T212 185
  configuration IMS T414 128, 132, **141**
  configuration IMS T800 71, 75, **84**
  configuration, external IMS T414 141, **142**
  configuration, external IMS T800 84, **85**
  configuration, internal IMS T414 **141**
  configuration, internal IMS T800 **84**
  data IMS T414 134
  data IMS T800 77
  direct access IMS T414 **150**
  direct access IMS T800 **93**
  dynamic IMS T414 132
  dynamic IMS T800 75
  external IMS T212 187, 190
  external IMS T414 130
  external IMS T800 73
  global 6
  IMS M212 216, 221
  IMS T212 187
  IMS T414 130
  IMS T800 73
  integrated 26
  interface 6, 20
  interface IMS T212 169, 185, **189**
  interface IMS T414 111, 128, **132**
  interface IMS T800 49, 71, **75**
  internal IMS T212 **187**, **190**

internal IMS T414 **130**, **133**
internal IMS T800 **73**, **76**
local 6
map 19, 21
map IMS T212 **188**
map IMS T414 **131**
map IMS T800 **74**
on-chip IMS T212 169
on-chip IMS T414 111
on-chip IMS T800 49
performance **303**
read IMS T414 136
read IMS T800 79
refresh IMS T414 128, 133, 142, 143, **147**, 148, 150
refresh IMS T800 71, 76, 85, 86, **90**, 91, 93
strobe IMS T212 **189**
strobe IMS T414 132, **134**, 137, 148, 150
strobe IMS T800 75, **77**, 80, 91, 93
wait IMS T212 190, **196**
wait IMS T414 132, 137, **148**, **149**
wait IMS T800 75, 80, **91**, **92**
MemReq
  IMS T212 **198**
  IMS T414 **150**
  IMS T800 **93**
MemStart 21
  IMS T212 183, 185, **187**
  IMS T414 126, 128, **130**
  IMS T800 69, 71, **73**
MemWait
  IMS T212 **196**
  IMS T414 **148**
  IMS T800 **91**
Message **6**, 8
  IMS T212 174
  IMS T414 116
  IMS T800 54
  pointer 33
  transfer 35
Microcode 29
  computing engine IMS T800 65
  cycle IMS T212 **189**
  cycle IMS T414 **132**
  cycle IMS T800 **75**
  scheduler 27, 31
  scheduler IMS T212 **172**
  scheduler IMS T414 **114**
  scheduler IMS T800 **52**
Microprocessor
  bus IMS C011 251
  bus IMS C012 275
  connection 27
  IMS C011 261
  IMS C012 283
MINUS **14**, 298
Mode 1
  IMS C011 251, **259**
  IMS M212 **217**
  link IMS C011 254, **259**

Mode 2
  IMS C011 251, **261**
  IMS M212 217, **218**
  link IMS C011 254, **261**
Modulo operator 298
MOVE2D 41, 302
Multiple length arithmetic **300**
Multiple processor 18
Multiplication performance **299**

Name 301
NaN
  IMS T800 **66**
Network 3, 5, 8, 16, 18, 21
  disk processor IMS M212 221
Node 4
NORMALISE 300
NOT **14**, 298
notCS
  IMS C011 **261**
  IMS C012 **283**
notMemCE
  IMS T212 190, 191, **193**, 196, 198
notMemRd
  IMS T414 **134**
  IMS T800 **77**
notMemRf
  IMS T414 **147**
  IMS T800 **90**
notMemS0-4
  IMS T414 **134**, 142, 143, 147, 148
  IMS T800 **77**, 85, 86, 90, 91
notMemWrB0-1
  IMS T212 190, **191**, 193, 195, 198
notMemWrB0-3
  IMS T414 **138**, 147
  IMS T800 **81**, 90

occam 3, 4, 8, 25, **26**
  channel 6, 19
  communication 8
  concurrency 8
  model **8**, 27
  process 3, 17, 18, 19
  program 3, 8, 16, 18
  synchronism 18
Operand 29
  IMS T212 169, 170, **176**
  IMS T414 111, 112, **118**
  IMS T800 49, 50, **56**
  register 30
  types **14**
Operating conditions
  IMS C004 242
  IMS C011 266
  IMS C012 288
  IMS T212 204
  IMS T414 156
  IMS T800 99

Operation
  arithmetic IMS T212   169
  arithmetic IMS T414   111
  arithmetic IMS T800   49
  code IMS T212   **176**
  code IMS T414   **118**
  code IMS T800   **56**
  logical IMS T212   169
  logical IMS T414   111
  logical IMS T800   49
Operator   **14**
  arithmetic   14, 298, 301
  bit   14, 298
  boolean   14, 298
  comparison   298, 301
  modulo   14, 298
  relational   14
  shift   14
Optimisation
  IMS T212   186
  IMS T414   129
  IMS T800   72
  program   17
OR   **14**, 298
Ordering details   345
Output   8, 9, 15, 21, 26, 298
  buffer IMS C011   264
  buffer IMS C012   286
  channel   26
  link IMS C004   **234**
  link IMS C011   **256**
  link IMS C012   **280**
  link IMS T212   185, **201**
  link IMS T414   128, **153**
  link IMS T800   71, **96**
  pins   20
  port IMS C011   **260**
  process   8, **9**, 11
  process IMS T212   174
  process IMS T414   116
  process IMS T800   54
  register IMS C011   **264**
  register IMS C012   **286**
Output Ready
  IMS C011   264, 265
  IMS C012   286, 287
OutputInt
  IMS C011   255, 264, **265**
  IMS C012   279, 286, **287**
Overflow
  stack IMS T212   169
  stack IMS T414   111
  stack IMS T800   49

Packet
  link   **36**
PAR   **10**, 12, 26, 299, 303
Parallel
  communication IMS C011   251
  communication IMS C012   275

construction   8, **10**, **16**, 26, 32
construction IMS T212   **173**
construction IMS T414   **115**
construction IMS T800   **53**
interface IMS C011   **261**
interface IMS C012   **283**
port IMS C011   **259**
process IMS T212   172
process IMS T414   114
process IMS T800   52
Part program   18
Pascal   19
Pattern recognition   31
Peek
  IMS T212   **185**
  IMS T414   **128**
  IMS T800   **71**
Performance   8, 25, 31, **297**
  bit counting   302
  bit reversal   302
  block move   302
  construction   303
  CRC   302
  Cyclic Redundancy Checking   302
  estimation   297
  external memory   **303**
  external RAM   303
  floating point   39, 297, **300**, **301**
  Floating point processor   38
  IMS T212   171
  IMS T414   113
  IMS T800   51
  integer   297
  interrupt latency   **304**
  link   303
  link IMS C004   234
  link IMS C011   256
  link IMS C012   280
  link IMS T212   201
  link IMS T414   153
  link IMS T800   96
  measurement   18
  multiple length arithmetic   **300**
  multiplication   **299**
  predefined maths   300
  priority   **304**
  product   **299**
  special purpose functions   **302**
  square root   302, 303
  TIMES   **299**
  wait states   303
Peripheral   21
  access   **15**
  control transputer   21
  device   26
  interface IMS M212   215
  memory mapping   21
Phase lock loop
  IMS C004   **231**
  IMS C011   **253**

IMS C012 **277**
IMS T212 **182**
IMS T414 **124**
IMS T800 **67**
Pipelined vector processor 39
PLACE 16
PLACED PAR 16
Placement **16**, 19
PLLx 341
  IMS T212 **189**
  IMS T414 **132**
  IMS T800 **75**
PLUS 14, 298
Pointer
  IMS T212 169
  IMS T414 111
  IMS T800 49
  instruction 32
  instruction IMS T212 173
  instruction IMS T414 115
  instruction IMS T800 53
  message 33
  workspace **30**
Poke
  IMS T212 **185**
  IMS T414 **128**
  IMS T800 **71**
Port **15**
  asynchronism 15
  disk IMS M212 215
  input IMS C011 **259**
  output IMS C011 **260**
  parallel IMS C011 **259**
  synchronism 15
PORT 21
Power 20
  IMS C004 **231**
  IMS C011 **253**
  IMS C012 **277**
  IMS T212 **182**
  IMS T414 **124**
  IMS T800 **67**
  rating IMS C004 **244**
  rating IMS C011 **269**
  rating IMS C012 **291**
  rating IMS T212 **207**
  rating IMS T414 **159**
  rating IMS T800 **102**
Prefetch 31
Prefix function **30**, 31
  IMS T212 **170**
  IMS T414 **112**
  IMS T800 **50**
PRI PAR 16
Primitive 298
Primitive type **13**
Priority 16
  bootstrap IMS T212 185
  bootstrap IMS T414 126
  bootstrap IMS T800 71

floating point IMS T800 **65**
IMS T212 **172**, **173**, 185
IMS T414 **114**, **115**, 128
IMS T800 **52**, **53**, 71
level 27, 28
performance **304**
timer IMS T212 **174**
timer IMS T414 **116**
timer IMS T800 **54**
PROC **14**, 18, 19, 298, 301
ProcClockOut
  IMS T212 **189**
  IMS T414 **132**
  IMS T800 **75**
Procedure 299
Procedures **14**
Process 4, 5, 8, 9, 10, **14**
  active/inactive 32
  active/inactive IMS T212 **172**, 173
  active/inactive IMS T414 **114**, 115
  active/inactive IMS T800 **52**, 53
  assignment 8, **9**
  channel 26
  communication IMS T212 **174**
  communication IMS T414 **116**
  communication IMS T800 **54**
  concurrent 10, 18, 26
  deschedule 35
  execution 32, 33, 34
  hardware **5**
  high priority IMS T212 186
  high priority IMS T414 129
  high priority IMS T800 72
  IMS T212 **172**, 186
  IMS T414 **114**, 129
  IMS T800 **52**, 72
  input 8, **9**, 11
  input IMS T212 **174**
  input IMS T414 116
  input IMS T800 54
  latency IMS T212 **173**
  latency IMS T414 **115**
  latency IMS T800 **53**
  list 32
  list IMS T212 **172**, 173
  list IMS T414 **114**, 115
  list IMS T800 **52**, 53
  low priority IMS T212 186
  low priority IMS T414 129
  low priority IMS T800 72
  mapping 8
  monitor 18
  new 32
  occam 18, 19
  output 8, **9**, 11
  output IMS T212 **174**
  output IMS T414 116
  output IMS T800 54
  parallel IMS T212 172
  parallel IMS T414 114

parallel IMS T800 52
primitive 26
queue IMS T212 **172**
queue IMS T414 **114**
queue IMS T800 **52**
reschedule 35
sequential IMS T212 169
sequential IMS T414 111
sequential IMS T800 49
simulation 18
software **4**
switch time 32
switch time IMS T212 **173**
switch time IMS T414 **115**
switch time IMS T800 **53**
timing IMS T212 **174**
timing IMS T414 **116**
timing IMS T800 **54**
Processor 20, 28
  clock IMS T212 **174**
  clock IMS T414 **116**
  clock IMS T800 **54**
  IMS M212 215, **217**
  IMS T212 **169**
  IMS T414 **111**
  IMS T800 **49**
  multiple 18
  speed IMS T800 47
  speed select IMS T800 **68**
ProcSpeedSelect0-2
  IMS T800 **68**
Product performance **299**
Program
  bootstrap IMS T212 199
  configuration **16**
  development **18**
  occam 18
  optimisation 17
  part 18
Programmable
  components **5**
  device 26
  i/o IMS C011 251
Programming 4
  model 8
  structure 30
protocol
  CHAN OF 298
Protocol
  link **36**

Q0-7
  IMS C011 255, 260
QAck
  IMS C011 260
Queue 33
  priority IMS T212 **172**
  priority IMS T414 **114**
  priority IMS T800 **52**
  process IMS T212 174
  process IMS T414 116
  process IMS T800 54
  timer IMS T212 **174**
  timer IMS T414 **116**
  timer IMS T800 **54**
QValid
  IMS C011 255, 260

RAM 303
  IMS T212 187
  IMS T414 126, 130
  IMS T800 69, 73
Read
  cycle IMS T212 **190**
  cycle IMS T414 **134**, 135
  cycle IMS T800 **77**, 78
  data IMS C011 **265**
  data IMS C012 **287**
  dynamic memory cycle IMS T414 136
  dynamic memory cycle IMS T800 79
  external cycle IMS T414 135, 136
  external cycle IMS T800 78, 79
REAL 301
Real time 18
REAL32 **13**, 301, 302
  conversion 301
REAL64 **13**, 301, 302
  conversion 301
Refresh
  memory IMS T414 128, 133, 142, 143, 146,
    147, 148, 150
  memory IMS T800 71, 76, 85, 86, 89, **90**, 91,
    93
Register
  A IMS T212 **169**, 170, 186
  A IMS T414 **111**, 112, 129
  A IMS T800 **49**, 50, 72
  B IMS T212 **169**
  B IMS T414 **111**
  B IMS T800 **49**
  C IMS T212 **169**
  C IMS T414 **111**
  C IMS T800 **49**
  CPU **28**, **29**
  data input IMS C011 **261**
  data input IMS C012 **283**
  FA IMS T800 **65**
  FB IMS T800 **65**
  FC IMS T800 **65**
  Floating point **28**
  I IMS T212 186
  I IMS T414 129
  I IMS T800 72
  IMS C011 **261**
  IMS C012 **283**
  IMS M212 215
  IMS T212 **169**, 172, 185
  IMS T414 **111**, 114, 128
  IMS T800 **49**, 52, 71
  input data IMS C011 261, 265

input data IMS C012 283, 287
input status IMS C011 261, **264**, 265
input status IMS C012 283, **286**, 287
link interface 35
operand 30
operand IMS T212 170
operand IMS T414 112
operand IMS T800 50
output data IMS C011 261, **264**, 265
output data IMS C012 283, **286**, 287
output status IMS C011 261, **264**, 265
output status IMS C012 283, **286**, 287
process list 32
timer IMS T212 **174**
timer IMS T414 **116**
timer IMS T800 **54**
W IMS T212 183
W IMS T414 126
W IMS T800 69
workspace IMS T212 170
workspace IMS T414 112
workspace IMS T800 50
REM 14, 298, 301
Repetition construction **12**
Replication construction **12**
Replication performance 299, 303
Reschedule **32**, 35, 37
Reset 20
  IMS C004 **233**, 247
  IMS C011 **255**
  IMS C012 **279**
  IMS T212 **183**, 185, 186
  IMS T414 **126**, 128, 129
  IMS T800 **69**, 71, 72
RnotW
  IMS C011 **261**
  IMS C012 **283**
ROM 21
  bootstrap code IMS T212 **187**
  bootstrap code IMS T414 **130**
  bootstrap code IMS T800 **73**
  IMS M212 216
  IMS T212 183
  IMS T414 126
  IMS T800 69
ROTATELEFT 300
ROTATERIGHT 300
RS0-1
  IMS C011 **261**
  IMS C012 **283**
Run time 19

SA400/450
  IMS M212 215, **217**
Scheduler 27, 31, **32**
  IMS T212 **172**, 173, 174
  IMS T414 **114**, 115, 116
  IMS T800 **52**, 53, 54
list 33
operation 32

SCSI
  bus IMS M212 221
  interface IMS M212 **222**
Selection construction **12**
Selector sequence
  floating point IMS T800 **56**, **65**
SeparateIQ
  IMS C011 **254**
SEQ 10, 11, 12, 26, 299, 303
Sequential
  construction 8, **10**, 11, 26
  process IMS T212 169
  process IMS T414 111
  process IMS T800 49
  processing 29
Serial
  data **36**
  protocol **36**
SHIFTLEFT 300
SHIFTRIGHT 300
Sieve of Erastosthenes 303
Silicon 28, 40, 43
Single byte instruction
  IMS T212 **170**
  IMS T414 **112**
  IMS T800 **50**
Skew
  strobe IMS T414 132, **138**
  strobe IMS T800 75, **81**
SKIP 298
Software
  IMS M212 218
  kernel 31
  kernel IMS T212 172
  kernel IMS T414 114
  kernel IMS T800 52
Special purpose functions **302**
Speed
  benchmark 38
  communication 27
  link 36
  processor 38
  processor IMS T800 47
  select IMS T800 **68**
SQRT 302
Square 14
Square root 302
  performance 303
ST506/412
  IMS M212 215, **217**
Stability
  clock IMS C004 231
  clock IMS C011 253
  clock IMS C012 277
  clock IMS T212 182
  clock IMS T414 124
  clock IMS T800 67
Stack
  evaluation 28, **29**, 33
  evaluation IMS T212 **169**, 173, 174

evaluation IMS T414 **111**, 115, 116
evaluation IMS T800 **49**, 53, 54
floating point IMS T800 **65**
optimise **38**
overflow 29, 38
overflow IMS T212 169
overflow IMS T414 111
overflow IMS T800 49
Start
  bit **36**
Status
  IMS T212 185
  IMS T414 128
  IMS T800 71
  register IMS C011 261
  register IMS C012 283
Stop
  bit **36**
STOP **17**, 298
Store
  instruction IMS T212 170
  instruction IMS T414 112
  instruction IMS T800 50
String **15**
Strobe
  memory IMS T212 **189**
  memory IMS T414 132, **134**, 137, 148, 150
  memory IMS T800 75, **77**, 80, 91, 93
  skew IMS T414 132, **138**
  skew IMS T800 75, **81**
  timing IMS T414 **137**
  timing IMS T800 **80**
  write IMS T414 **138**
  write IMS T800 **81**
Structure
  data IMS T212 170
  data IMS T414 112
  data IMS T800 50
Subscript **13**
  constant 298
  expression 298
  variable 298
Synchronisation
  link IMS C004 234
  link IMS C011 256
  link IMS C012 280
  link IMS T212 201
  link IMS T414 153
  link IMS T800 96
  point IMS T800 **65**
System services 20
  IMS C004 **231**
  IMS C011 **253**
  IMS C012 **277**
  IMS T212 **182**
  IMS T414 **124**
  IMS T800 **67**

testerr
  IMS T212 186

  IMS T414 129
  IMS T800 72
tim 15
Time **15**
  delay IMS T212 172
  delay IMS T414 114
  delay IMS T800 52
  process switch IMS T212 **173**
  process switch IMS T414 **115**
  process switch IMS T800 **53**
  real 18
  slice IMS T212 **172**, 173
  slice IMS T414 **114**, 115
  slice IMS T800 **52**, 53
  slice period IMS T212 **173**
  slice period IMS T414 **115**
  slice period IMS T800 **53**
Timeout 17
Timer 13, **15**, 18, **37**, 299
  AFTER 299
  clock 15
  clock IMS T212 **174**
  clock IMS T414 **116**
  clock IMS T800 **54**
  delay 15
  IMS T212 173
  IMS T414 115
  IMS T800 53
  input 299
  processor 37
  queue IMS T212 **174**
  queue IMS T414 **116**
  queue IMS T800 **54**
  register IMS T212 **174**
  register IMS T414 **116**
  register IMS T800 **54**
TIMER **13**
TIMES **14**, 298
  performance **299**
Timing 8
  strobe IMS T414 **137**
  strobe IMS T800 **80**
Tm
  IMS T414 **132**
  IMS T800 **75**
TPtrLoc1, TPtrLoc2
  IMS T212 **187**
  IMS T414 **130**
  IMS T800 **73**
Transcendental function 39
Transfer message 35
Transmission link 36
Transputer
  array IMS M212 221
  clock 7
  development system 8, 16, 17, 21
  development system IMS T414 132
  development system IMS T800 75
  family 27
  interconnection 28

products 27
TRUE 14
Tstate
  IMS T212 **190**
  IMS T414 **132**, 134, 142
  IMS T800 **75**, 77, 85
TTL compatibility
  link IMS C004 234
  link IMS C011 256
  link IMS C012 280
  link IMS T212 201
  link IMS T414 153
  link IMS T800 96
Type **13**
  array **13**
  BOOL **13**
  BYTE **13**
  CHAN OF **13**
  floating point **13**
  INT **13**
  INT16 **13**
  INT32 **13**
  INT64 **13**
  primitive **13**
  REAL32 **13**
  REAL64 **13**
  record **13**
  TIMER **13**
  variant **13**

Value
  constant **30**
  data 29
  literal **30**
Variable 8, **9**, 13, 14, 26, 298, 301
  array 298
  IMS T212 170
  IMS T414 112
  IMS T800 50
  subscript 298
  temporary IMS T212 171
  temporary IMS T414 113
  temporary IMS T800 51
VCC 20
  IMS C004 **231**
  IMS C011 **253**
  IMS C012 **277**
  IMS T212 **182**
  IMS T414 **124**
  IMS T800 **67**
VLSI 26, 43

Wait
  IMS T212 **196**
  IMS T414 132, 137, 143, **148**, **149**
  IMS T800 75, 80, 86, **91**, **92**
  state generator IMS T212 **196**
  state IMS T212 190
Whetstone benchmark 38, 39
WHILE **12**, 26, 299

Word
  access IMS T212 **195**
  length 28, 29, 31
Workspace 32, 34
  disk IMS M212 217
  IMS T212 169, 173, 185
  IMS T414 111, 115, 126
  IMS T800 49, 53, 71
  instruction IMS T212 173
  instruction IMS T414 115
  instruction IMS T800 53
  pointer **30**
  register IMS T212 170
  register IMS T414 112
  register IMS T800 50
Write
  cycle IMS T212 **190**, **191**
  cycle IMS T414 134, **138**
  cycle IMS T800 77, **81**
  data IMS C011 **265**
  data IMS C012 **287**
  early IMS T414 142
  early IMS T800 85
  late IMS T414 142
  late IMS T800 85
  strobe IMS T212 **191**
  strobe IMS T414 **138**
  strobe IMS T800 **81**

Appendix E

ordering

E   Ordering

The following tables indicate the designation of speed and package selections for the various devices. Speed of **ClockIn** is 5MHz for all parts. For transputers, processor cycle time is nominal; it can be calculated more exactly using the phase lock loop factor **PLLx**, as detailed in the external memory sections in the Family Characteristics part.

Table E.1: IMS T212 ordering details

INMOS designation	Instruction throughput	Processor clock speed	Processor cycle time	PLLx	Package
IMS T212A-G17S	8.75 MIPS	17.5 MHz	57 ns	3.5	Ceramic Pin Grid
IMS T212A-G20S	10.00 MIPS	20.0 MHz	50 ns	4.0	Ceramic Pin Grid
IMS T212A-J17S	8.75 MIPS	17.5 MHz	57 ns	3.5	Plastic J-Bend
IMS T212A-J20S	10.00 MIPS	20.0 MHz	50 ns	4.0	Plastic J-Bend

Table E.2: IMS T414 ordering details

INMOS designation	Instruction throughput	Processor clock speed	Processor cycle time	PLLx	Package
IMS T414B-G15S	7.5 MIPS	15 MHz	67 ns	3.0	Ceramic Pin Grid
IMS T414B-G20S	10.0 MIPS	20 MHz	50 ns	4.0	Ceramic Pin Grid
IMS T414B-J15S	7.5 MIPS	15 MHz	67 ns	3.0	Plastic PLCC J-Bend
IMS T414B-J20S	10.0 MIPS	20 MHz	50 ns	4.0	Plastic PLCC J-Bend

Table E.3: IMS T800 ordering details

INMOS designation	Instruction throughput	Processor clock speed	Processor cycle time	PLLx	Package
IMS T800C-G17S	8.75 MIPS	17.5 MHz	57 ns	3.5	Ceramic Pin Grid
IMS T800C-G20S	10.00 MIPS	20.0 MHz	50 ns	4.0	Ceramic Pin Grid
IMS T800C-G30S	15.00 MIPS	30.0 MHz	33 ns	6.0	Ceramic Pin Grid

Table E.4: IMS M212 ordering details

INMOS designation	Instruction throughput	Processor clock speed	Processor cycle time	PLLx	Package
IMS M212B-G15S	7.5 MIPS	15 MHz	67 ns	3.0	Ceramic Pin Grid
IMS M212B-G20S	10.0 MIPS	20 MHz	50 ns	4.0	Ceramic Pin Grid
IMS M212B-J15S	7.5 MIPS	15 MHz	67 ns	3.0	Plastic PLCC J-Bend
IMS M212B-J20S	10.0 MIPS	20 MHz	50 ns	4.0	Plastic PLCC J-Bend

Table E.5: IMS C004 ordering details

INMOS designation	Package
IMS C004B-G20S	Ceramic Pin Grid Array

Table E.6: IMS C011 ordering details

INMOS designation	Package
IMS C011A-P20S	28 pin plastic dual-in-line
IMS C011A-S20S	28 pin ceramic sidebraze

Table E.7: IMS C012 ordering details

INMOS designation	Package
IMS C012A-P20S	24 pin plastic dual-in-line